The Birth of Stars and Planets

Star formation is the fundamental cosmic process which makes galaxies visible and regulates the evolution of normal matter in the Universe. Stars forge heavier elements from the primordial hydrogen and helium that emerged from the Big Bang. While short-lived massive stars generate intense radiation, die in cataclysmic supernova explosions, and regulate the behavior of surrounding star-forming clouds, planets that formed around low-mass stars provide the long-lived stable environments required for the evolution of life. We must look to the stars to understand our origins.

New instruments and technologies are now enabling the exploration of our origins. Scientists are beginning to understand the beauty and complexity of star and planet formation and its role in cosmic evolution. This book combines the latest astronomical images and data with descriptions of the exciting recent developments in the study of star and planet formation. The authors discuss isolated star birth in dark clouds, the formation of star clusters and nebulae, the "ecology" of interstellar gas and dust, and the violent starbursts that may produce black holes. They relate these processes to the evolution of galaxies and the origin of life on Earth. Written in a non-technical language, the book will appeal to readers with an interest in understanding the Universe from its beginnings to life as we know it today.

JOHN BALLY is a professor in the Department of Astrophysical and Planetary Sciences at the University of Colorado, Boulder. He studied at the University of California at Berkeley and obtained his Ph.D. at the University of Massachusetts in Amherst in 1980. Between 1980 and 1991, he was a researcher in the Radio Physics Research Laboratory at the AT&T Bell Laboratories in Holmdel, New Jersey. He moved to the University of Colorado in 1991.

BO REIPURTH is a professor at the Institute for Astronomy, University of Hawaii. He obtained his Ph.D. in 1981 at the University of Copenhagen and worked there until moving to the European Southern Observatory in Chile for 11 years. He then spent 4 years at the University of Colorado before joining the faculty at the University of Hawaii in 2001.

The Birth of Stars and Planets

John Bally
University of Colorado, Boulder

Bo Reipurth
University of Hawaii

CAMBRIDGE
UNIVERSITY PRESS

CAMBRIDGE UNIVERSITY PRESS
Cambridge, New York, Melbourne, Madrid, Cape Town, Singapore, São Paulo

Cambridge University Press
The Edinburgh Building, Cambridge CB2 2RU, UK

Published in the United States of America by Cambridge University Press, New York

www.cambridge.org
Information on this title: www.cambridge.org/9780521801058

First published 2006

Printed in China by Imago

A catalog record for this publication is available from the British Library

ISBN-13   978-0-521-80105-8 hardback
ISBN-10   0-521-80105-2 hardback

Contents

Part II Planetary systems 141

Part III The cosmic context

Foreword

Stars and galaxies are fundamental units of our visible Universe. The question of their origins is among the most profound issues contemplated by astronomers in the past as well as today. Whereas the formation of galaxies is a problem for which we are only beginning to glimpse the outline of a solution, the birth of stars and their accompanying planets is rapidly being understood. Technical developments during the last twenty-five years have opened up wavelength regions that have increasingly allowed us to probe into the interior of dark clouds to witness directly the gestation and birth of stars. New high resolution observing techniques are being developed that permit us to explore the disks out of which planets are formed. It is possible – even likely – that within the next decade all the major pieces of a full picture of star and planet formation will be in place.

With this book we have wanted to share with a wider audience the numerous exciting developments and intellectual milestones on the road to an understanding of our origins. In writing this book, we had to make a set of choices. First, we have wanted to make the book accessible to people who do not have a specific background in astronomy and physics, but who are willing to follow a complicated argument. We have therefore written extensive notes to give support when physical, technical, or astrophysical concepts inevitably had to be drawn into the presentation. Also, we have chosen to include equations only in the footnotes, not in the main text, where we have instead relied on a discussion of physical principles. Second, although this book is not meant as a pretty coffee-table book, we have wished to share the amazing beauty of star and planet formation, so we have made efforts to select many of the most stunning images of the field. In order to strengthen this aesthetic aspect, we have chosen to include only a few diagrams and to place those mostly in the note section. Finally, in a subject as large as this one, inevitably not every aspect can be covered in equal detail. The issues presented in the book are therefore colored by our own views and particular research interests.

Most of the images presented in this book show far more detail, color, and texture than any human eye could see, even through the largest telescope. And some images show objects that emit their 'light' at wavelengths to which our eyes are totally blind. The use of color in the presentation of astronomical images therefore differs from the conventions of ordinary photography.

Most electronic image sensors are color-blind in the sense that they merely record the intensity of light over some range of wavelengths to which they are sensitive. To obtain a color image, a set of exposures obtained through several color filters, for example blue, green, and red, must be combined. The result is a 'natural-color' image. Many astronomical images are recorded with filters and detectors which transmit and record wavelengths radically different from what our eyes can see. For example, an image can be recorded at radio wavelengths, a second at infrared wavelengths, and a third in the ultraviolet. Such images cannot be presented at the same 'color' as they were recorded. In this case, the most natural solution is to present the long wavelength image as red, the medium wavelength image as green, and the short wavelength image as blue. Although the resulting color image is not 'true-color', the color is at least presented in wavelength order. Another common use of color is to present data taken through specific filters which only transmit narrow ranges of wavelengths corresponding to particular atoms, ions, or molecules. In the chapters presenting images of jets, outflows, and nebulae, many were recorded using such narrow-band filters which transmit the spectral lines of atomic hydrogen, ionized sulphur, or some other species. In these cases, color is merely used to identify regions with different physical conditions. Despite these limitations, we believe the resulting images convey the beauty of astronomical objects.

We are thankful to our colleagues Alan Boss, George Herbig, David Jewitt, Sarah Knights, Nathan Smith, and Henry Throop who commented on some or all of the chapters at various stages of their development. Our appreciation also goes to the numerous colleagues who provided us with copies of their images: J. Alves, P. Armitage, C. Aspin, Matthew Bate, M. Bissell, H. Boffin, Alan Boss, K. Bucka-Lassen, J.-C. Cuillandre, Tony and Daphne Dallas, T. Dame, Dan Durda, Jack Eastman, J. English, B. Hansen, P. Hartigan, R.P. Harvey, George Herbig, J.J. Hester, D. Johnstone, W. Kley, C.J. Lada, Dante Lauretta (Image from 'A Color Atlas of Meteorites in Thin Section', D.S. Lauretta and M. Killgore), M. Liu, M. McCaughrean, J.W. McNeil, N. Moeckel, Ben Moore, L. Stougaard Nielsen, R. O'Dell, D. Padgett, Thomas Preibisch, T.A. Rector, L.F. Rodriguez, Johannes Schedler, Stefan Seip (www.astromeeting.de), Nathan Smith, R.S. Sutherland, R. Taylor, David Thompson, J. Walawender, Wei-Hao Wang, and A. Whitworth. We are also grateful to the organisations ESO, Gemini, NASA, NOAO, NRAO, and Subaru for permission to reproduce images.

We thank Jacqueline Garget and Cambridge University Press for great patience as we passed multiple deadlines for submission of this manuscript.

Finally, but not least, heartfelt thanks to our wives, Kim and Mercia, for accepting too many long nights of writing.

Part I Stars and clusters

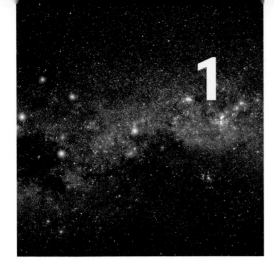

1 Our cosmic backyard

The big questions

What are stars? Why do they shine? How did the Earth, the Sun, the solar system, and stars form? How common are planets around other stars? How did life on Earth arise? Is life common in the Universe? What is the history of our planet and the life it supports? These, and other similar questions, have stimulated thinking, discussion, and heated debate during all of recorded history. We are entering an era in which scientific research is starting to provide answers.

Humans are curious about their environment and seek to understand the Universe. Every culture has constructed a "world view": a description of their reality. From ancient nomadic tribes to the cultures of Egypt, Greece, and Rome, the great civilizations of China and India, the inhabitants of the New World and the remotest islands in the Pacific, curious humans have sought to understand Nature. Virtually all of these "world views" were based on myth, religion, or philosophical discourse.

Within the last half millennium, common experience, observation, and the results of experiments have become increasingly incorporated into the description of the world around us. As the Dark Ages came to a close in Europe, the Renaissance gave birth to the development of modern science. Building on the astronomical traditions of the Arab world, and motivated by observations of the Sun, Moon, and planets and by the behavior of objects on Earth, early European practitioners of science developed mathematical descriptions of the mechanics of motion. Isaac Newton related the movement of the Moon to the behavior of falling bodies through his theory of gravitation. With Newton's theory came a general understanding of the motions of bodies in our solar system.

The revolution in science that started with Galileo, Kepler, and Newton produced the intellectual underpinnings of the Industrial Revolution that would ultimately lead to our current way of life. It also initiated a profound series of discoveries in astronomy that would lead to our present view of the Universe.

The invention of the telescope during the first decade of the seventeenth century created an opportunity to see fainter objects and finer details on celestial bodies than visible to the eye alone. These early telescopes were used to scrutinize the Sun, Moon, and planets; they led to the discovery of the phases of Venus and Mercury, the moons of Jupiter and Saturn, and to the study of the mysterious comets. As telescopes became more powerful, a new planet, Uranus, was discovered in 1781. And, as the motions of bodies within our Solar System became better understood, discrepancies between the observed and predicted movements of this new planet led to the *prediction* of yet another large body. In a spectacular demonstration of Newton's theory of gravity and its application to celestial motions, the planet Neptune was discovered in 1846 close to its predicted location.

With telescopes of ever increasing power, astronomers learned that there are vastly more stars in the sky than our unaided eyes alone can see, and that these stars are very far away. Furthermore, here and there between the stars, astronomers found faint glowing clouds, "nebulae", whose nature was not understood until the twentieth century.

The Industrial Revolution provided new tools to further the progress of astronomy. Until the middle of the nineteenth century, all astronomical observations through telescopes were made with the human eye and our perception of the Universe was limited by our eyesight. Observations had to be recorded by hand either as sketches or as numbers corresponding to the brightness, position, size, or other measurable characteristics of the object under study. Furthermore, the eye cannot accumulate light like a photographic time-exposure. When fully adapted to the dark, our eyes respond only to a few percent of the light that falls on the retina and the "exposure time" of our eyes is only about one-thirtieth of a second.

The light sensitive pigments in our eyes only respond to a very restricted set of wavelengths. During the day when there is plenty of light, our eyes employ color sensitive cells called cones. But cones are not very sensitive. To sense faint light at night, the eye uses rods. However, rods have essentially no color sensitivity. The brightest stars in the sky are just at the detection limit of the color sensitive cone cells. Therefore some people, especially the young, can recognize colors of the very brightest stars.

Chemistry led to a new way of "seeing." In the second half of the nineteenth century, photography was invented. Emulsions of silver compounds deposited on glass plates provided permanent records of the objects under study. Furthermore, these films could accumulate incident light, making objects visible that are too dim to see by eye, even when aided by a telescope. By taking exposures lasting many hours, stars and nebulae many times fainter than what the eye could see could be recorded (Figure 1.1).

Astronomers also adapted another tool of the chemists to the study of celestial phenomena: the "spectroscope."[1] By passing the light of a celestial object through a prism or diffraction grating, it could be sorted into its constituent colors. Then, by recording the patterns of light produced by a spectroscope on photographic emulsions or with other types of light recording device, a "spectrograph" could be used to analyze the temperatures, motions, and chemical makeup of astronomical objects. Such analyses

Figure 1.1. A wide-angle view towards the center of our own Milky Way galaxy. (Wei-Hao Wang).

permitted the identification of the substances that either emitted or absorbed light. The marriage of photography and spectroscopy led to the study of the physical and chemical properties of celestial objects: the subject that is today called "astrophysics."

Astrophysical measurements revealed that stars are mostly made from the light chemical elements hydrogen and helium. About 70 percent of the mass of most stars consists of hydrogen, about 28 percent consists of helium. All of the other 90 known stable elements make up a mere 1 to 2 percent of the mass of a typical star. This is in sharp contrast to the Earth, which is made mostly of heavier elements such as iron, silicon, aluminum, and oxygen.

Stars: beacons and building blocks

Stars are the visible constituents of the Universe and play a role in astronomy similar to that of atoms in chemistry. Stars produce the light and energy that stirs and heats the gas and dust in space and makes life on planets like Earth possible. Furthermore, stars are element factories. Most chemical elements heavier than helium that exist in the Universe today were forged by thermonuclear fusion reactions in stellar interiors from the hydrogen and helium that emerged from the Big Bang. Indeed, life, and most of the Earth, is made of recycled stardust expelled from the hearts of dying stars. Earth, and life as we know it, could not exist without the thermonuclear burning of elements in stellar cores.

Stars come in a range of sizes, masses and temperatures. While some are hotter and more massive than the Sun, most stars are slightly cooler and smaller.

To be a star, an object has to have at least about one tenth the mass of the Sun. Any self-gravitating cloud that attempts to become a star with less material would never develop the pressures and temperatures in its interior needed for the ignition of the nuclear processes that make stars shine. Objects with less mass do exist. However, they cannot sustain a steady light for long periods of time. They may glow like normal stars for a few million years after birth by tapping into the gravitational energy released by their formation and contraction. But, since they cannot tap any long-lasting source of energy in their cores, they shrink and over billions of years fade to near invisibility. Such failed stars are known as *brown dwarfs*. Old brown dwarfs that have cooled come to resemble the gas giant planets such as Jupiter.

The most massive stars are about 100 times heavier than the Sun and burn with a brilliance more than a million times greater. Consequently, they consume all available fuel in only a few million years and die in cataclysmic explosions. Even more massive stars apparently do not exist. Stars containing more than roughly 100 solar masses would produce such enormous amounts of energy in their cores that the pressure of their own light would blow their surface layers into space. The furious rate of energy generation would disrupt such a would-be super-massive star.[2] As we shall see in Chapter 9, the pressure of starlight also affects the evolution of the most massive stars that do exist. Over the course of their relatively brief but brilliant lives, massive stars blow powerful *stellar winds* into their environment. Such stellar winds can remove most of a massive star's mass by the time it exhausts its nuclear fuel and dies.

While the Sun will live for about 10 *billion* years from the time of its formation, massive stars consume their supply of hydrogen fuel in a mere few *million* years. Thus, all massive stars must be young. For such stars to exist in the sky today, the process of star birth must be an on-going process.

Star formation determines the evolution of ordinary matter[3] in the Universe and is therefore a fundamental cosmic process. Furthermore, planets appear to originate as a direct by-product of star formation. To understand our origins, we must therefore look to the stars. We must understand how stars generate light and energy, how they create the chemical elements, how they die and recycle their contents and, most importantly, how the process of star and planet formation works. This latter subject is the topic of this book.

Stars are very distant suns. If we assume that stars are about as luminous as our Sun, their distances can be estimated from a comparison of their apparent brightness compared to the Sun. This crude estimate implies that even the nearest stars are hundreds of thousands of times farther from us than the Sun. The Earth–Sun separation is the unit of distance astronomers call the Astronomical Unit, abbreviated as AU[4].

Careful measurements of the apparent positions of the nearest stars over the course of a year confirmed this conjecture by the middle of the eighteenth century. As the Earth moves around the Sun, the positions of nearby stars on the sky change ever so slightly with respect to more distant ones. Painstaking measurements of this "parallax" is still the most reliable method for the determination of stellar distances.

Why do stars shine? What makes starlight? For how long do stars live? The starlight we see is emitted by hot gases at stellar surfaces. As these gases radiate into space, they cool. To shine continuously, stellar surfaces must remain hot, and energy must rise from the depths of the stellar interior to replenish what is lost to space in the form of starlight.

The energy source that enables stars like our Sun to shine for billions of years remained unknown until the middle of the twentieth century. Calculations performed during the nineteenth century indicated that if the Sun shone by consuming chemical fuel, such as a burning pile of coal, it would stop shining within less than a hundred thousand years. On the other hand, the Sun might conceivably derive its energy from gravity. Just as a hydroelectric power plant uses descending streams of water to generate electricity, the Sun could extract energy from gravity by contracting. If the Sun could shrink, gravitational energy would be released. To replenish the energy radiated into space as sunlight, the Sun would have to shrink by more than a factor of two in just a million years. Yet, by the beginning of the twentieth century, geologists found evidence that many rocks on Earth have ages of hundreds of millions of years. They concluded that the Sun has shone for far longer than either chemical or gravitational energy would allow. The Sun's source of power remained a mystery until the 1930s when the nuclear forces and nuclear energy were discovered.

How ironic it is that to understand the energy source of stars, among the largest recognizable bodies in the Universe, we have to peer deep into the microscopic realm of atoms. By the end of the nineteenth century and beginning of the twentieth we learned that atoms consist of low mass, electrically charged electrons buzzing about a much denser and a thousands of times more massive atomic nucleus.[5] These nuclei, which are minutely small compared to the already minuscule atom,[6] contain positively charged protons and uncharged neutrons. But, the electron's charge is negative, and opposite charges attract. Therefore, the electrons become trapped by the electrical attraction of the positively charged protons.

The simplest atom, hydrogen, consists of a single proton orbited by a single electron. All other elements contain two or more protons in their nuclei surrounded by a cloud of an identical number of electrons. Each time a proton is added to a nucleus (usually requiring also the addition of one or more neutrons), the atomic core of a new chemical element is produced. The additional electric charge of the resulting nuclei can bind new electrons to form electrically neutral atoms, each with distinct chemical and physical properties. All told, there are 92 distinct stable elements in the Universe.

Why do atomic nuclei contain neutrons? While unlike charges attract, like charges repel. The enormous electric repulsion of the protons bound in atomic nuclei heavier than hydrogen must be overcome by a force that is much stronger than electricity. This force is carried by the neutrons, which serve to glue together the mutually repulsive protons in nuclei heavier than hydrogen. This "glue" is called the "strong nuclear force."[7] Its binding energy can be released when heavy nuclei such as uranium are broken into lighter elements in a nuclear "fission" reaction which is utilized in all current nuclear power reactors. But even more energy can be released when light nuclei such as hydrogen combine to form heavier nuclei such as helium in a nuclear "fusion" reaction.[8]

By 1948, it was realized that it is the thermonuclear fusion of hydrogen into helium that fuels the Sun and most other stars. The nuclear fusion of hydrogen into helium can sustain the Sun's prodigious energy output for about 10 billion (10^{10}) years, more than a factor of two longer than the current 4.5 billion year age of the Solar system.

Stars have colors related to the temperatures at their surfaces which again are related to their masses and state of evolution. The Sun, with a surface temperature of about $6000°C$,[9] is yellow. Stars hotter than the Sun are bluer, while cooler stars are redder. But an additional pattern was discovered during the first decades of the twentieth century. The majority of blue stars are more luminous and more massive than the Sun. On the other hand, most red stars are less luminous and less massive. These "normal" stars, which make up a large majority (about 90 percent) of all stars, are burning hydrogen into helium in a thermonuclear fusion reaction.

The main exceptions to these "normal" stars are red giants and supergiants. As they exhaust their fuel of hydrogen, the stellar cores become denser and hotter. At the same time, the outer layers of such stars swell and cool. Therefore, as they evolve towards stellar death, stars become more luminous and red. As discussed in later chapters, the more massive stars may extend their lives a bit longer by burning the heavy elements in additional fusion reactions during a supergiant stage.

During the twentieth century, astronomers understood the structure and lives of stars. By the 1950s, they learned how stars generate energy through nuclear processes. In the 1960s they clarified how stars evolve. In the 1980s they began to understand stellar old age and death. In the 1990s they took major steps towards understanding how stars are born. This book deals with the brief period during which a star is born and the surprisingly violent processes associated with stellar genesis.

A 14-billion-year-old expanding Universe

As astronomers came to understand the nature of stars during the early decades of the twentieth century, they also uncovered unexpected features of the Universe. When we peer into the depths of space beyond our Milky Way with giant telescopes, we see that the Milky Way is only one of a vast number of galaxies, each containing billions of stars. Some galaxies are flattened and rotating systems of stars and gas with a pronounced spiral pattern. Others are elliptical or spherical collections of stars that contain relatively little gas. Still others have irregular shapes. Galaxies are huge. Moving with the speed of light (300 000 kilometers per second), it would take about one hundred thousand years to cross the dimensions of a typical galaxy. Therefore, these gigantic systems are said to be about 100 000 light years in diameter.

Galaxies tend to cluster in groups containing anywhere from tens to thousands of members, and form an enormous filamentary tapestry spread through space like gigantic cobwebs. Most of the light we see from these structures originates in stars.

By the late 1920s, the astronomer Edwin Hubble found evidence that the galaxies are moving away from each other. Their recession speeds are directly proportional to their distances; a galaxy twice as far from us moves twice as fast.

From the rate of expansion, the "age" of the Universe can be determined. But, what does this "age" mean? The galaxies are moving away from each other as if attached to the surface of an inflating balloon.[10] For how long has this hypothetical balloon been expanding? The answer is the "age" of the Universe. Within the last decade, the expansion rate of the Universe has been measured to a precision of a few percent, and the measurements indicate that the Universe is about 13.7 billion years old (give or take a few hundred million years).

The matter within the Universe becomes more tenuous and cooler as it expands. Therefore, the Universe was denser and hotter in the past. Direct evidence for much denser and hotter conditions in the distant past was uncovered by the middle of the twentieth century. The abundance of helium in the cosmos is far greater than the amount that could have been produced by stars by the nuclear fusion reactions that make them shine. In 1948, Ralph Alpher, Hans Bethe, and George Gamov proposed that this excess helium abundance was produced during the first few minutes following a spectacular explosion which created the Universe (and later was called the Big Bang), while the average temperature was still as hot as the center of our Sun today. This theory of helium production predicted that a relic of this hot past ought to be visible today as a faint glow of microwave radiation emanating from every direction on the sky.

In 1965, just such a glow of radio wavelength energy was discovered by Arno Penzias and Robert Wilson at the Bell Laboratories in New Jersey, USA. This radiation, which came to be known as the "cosmic microwave background" is light emitted by the expanding Universe when it was only about 300 000 years old. To understand this remarkable discovery, we have to recall that light travels at a finite speed.

When we see distant objects, we see them *not* as they are now, but as they were when the light we see left them. As we look at ever more distant galaxies with our telescopes, we look ever further back in time. Telescopes are time machines. Thus, we see the Moon as it was about a second before we see it. When we look at the Sun, we see it the way it was about 8 minutes ago. When we look at the nearest star visible to the naked eye, Alpha Centauri (Figure 1.2), we see it the way it was about four years ago. When we look at the Orion Nebula located 1500 light years distant, we see it the way it was when the Roman Empire fell. As we use our infrared telescopes to peer at the center of our Milky Way, we see it as it was about 30 000 years ago when the Earth was in the middle of the last ice age. The most distant object visible to the unaided eye is the Andromeda galaxy, the nearest system having a size similar to our Milky Way galaxy. When you see its dim glow, you are collecting light that has been traveling for about 2 million years! When we look at the most distant objects in the deepest images returned by the Hubble Space Telescope, we see across vast eons of time.

Deep images obtained with our most powerful instruments depict objects *billions* of light years from us, so these pictures show the Universe as it was billions of years ago. Long time-exposure images that penetrate far into cosmos, such as the "Hubble Deep Field" (Figure 1.3) provide us with glimpses of galaxies which are more than 10 billion years old. We can look back over vast timespans to see galaxies no older than 10 percent of the current age of our Universe! When we compare their appearance to closer

Figure 1.2. The southern Milky Way showing the nearest star Alpha Centauri (bright yellow star at the left) and the Southern Cross (middle). (Wei-Hao Wang).

galaxies, we can *see* the evolution of the Universe unfold over much of its 14 billion year history. Indeed, in astronomy, we are in a unique situation. Owing to the finite speed of light, we can actually watch the history of our Universe. All we have to do is record the light produced by objects at different distances; the greater the distance, the farther back in time we see.

But these faint and distant galaxies are *not* the most remote objects we can see. The cosmic microwave background provides us with a glimpse of matter in the Universe when it was a mere 0.003 percent of its current age (Figure 1.4). Pictures of the cosmic microwave background offer information about the conditions in the Universe when it was just 300 000 years old.

In these ancient times, there were no planets, stars, or galaxies. Instead, the sky was filled with the intense light emitted by a hot sea of hydrogen, helium, and trace amounts of a few light elements. In today's Universe,

Figure 1.3. The "Hubble Ultra Deep Field" showing
a field of extremely distant and faint galaxies in the
constellation Fornax located below Orion.
(NASA/STScI).

Figure 1.4. A picture of the cosmic microwave background showing the nearly uniform glow of hot plasma in the Universe emitted only about 300 000 years after the Big Bang. The image is based on data obtained with the Wilkinson Microwave Anisotropy Mapper (WMAP/NASA).

gravity has caused most of this matter to condense into gas clouds, stars, and galaxies. But in the young Universe, the hot gas was spread around very smoothly; its density was the same everywhere to about one part in a hundred thousand. And it was hot. So hot that hydrogen nuclei could not hold on to their electrons. The hydrogen atoms were *ionized*.[11] The mixture of protons, electrons, and helium atoms had a temperature of about 3000° C, half the current temperature of the Sun's surface.

Imagine floating in space when the Universe was younger than 300 000 years. There would be no stars in the sky. As you looked around, you would see a nearly uniform glow of red and near-infrared light coming at you from every direction. Most of the light would have wavelengths several times longer than those that your eyes can see. From every direction, the sky would radiate about one-sixteenth as much energy as the surface of the Sun. The Universe was very different in the distant past.

We can still see some of this light coming toward us from the furthest reaches of the ancient Universe. Soon after reaching an age of 300 000 years, the expanding Universe cooled sufficiently for hydrogen nuclei (protons) to combine with electrons to form the very first stable hydrogen atoms. Up to this time, the hydrogen gas was opaque. As on a foggy night, light could not travel far. But as hydrogen atoms condensed from the sea of protons and electrons, the "fog" became transparent. And the intense glow of the young Universe was allowed to travel through space unimpeded. Immediately after the condensation of the first atoms, our hypothetical cosmic traveler would see a sphere of clearing recede in all directions at nearly the speed of light.

Zipping forward 14 billion years to the present, we continue to "see" the wall of clearing moving away. This surface, which is still glowing with the light of the ancient hydrogen, appears to us to be nearly 14 billion light years distant. The wavelengths of light we see today have been stretched by the expansion of the Universe a thousand-fold into the millimeter wavelength portion of the spectrum.

From our vantage point, the cosmic microwave background looks like the dimly glowing edge of the visible Universe, racing away at nearly light speed. It effectively forms a horizon beyond which we cannot see. Our view is blocked by a distant wall of luminous fog formed by opaque hydrogen and helium ions.

The nearly uniform distribution of hydrogen and helium gas that emerged from the Universe immediately after the production of the cosmic microwave background was gradually transformed into stars, clusters of stars, and galaxies by the process of star formation.

In this book, we will first explore star formation in our own backyard: the corner of the Milky Way that we inhabit. As we develop the modern picture of star formation, we will shift our gaze to the larger scales of galaxies, and towards the end of the book, apply our understanding to the very early Universe. We will end by discussing the conditions under which the very first stars in the Universe formed and how star formation was responsible for the conversion of the smooth sea of atoms that emerged from the Big Bang into the star and galaxy-studded sky that surrounds us. In the process, we will gain insights into our own origins.

Astronomy is a monument to our understanding of Nature. An understanding which has provided the intellectual fuel to power the economy and the technology on which it is based. But ultimately, it is human curiosity that is the key to unlocking the secrets of the world. We are curious. We want to understand our place in the Universe. Astronomy has been a great motivator.

In the next chapter, we will discuss in more detail the tools that scientists use to investigate the nature of astronomical sources.

2 Looking up at the night sky

The golden age of astronomy

We live in a remarkable time. During the twentieth century, we developed telescopes that can "see" all of the wavelengths contained in the spectrum. We are the first generation to have the ability to observe the Universe in its full splendor.

Two key technologies have contributed to our ability to explore the Universe. First, with rockets, we gained access to space. Using telescopes flying above the obscuring veil of the Earth's atmosphere, we can see the entire spectrum of "light" from long radio wavelengths to the ultra-short gamma rays.[1] (In order of decreasing wavelength, the main portions of this spectrum include the radio, infrared, the visual, ultraviolet, X-ray, and gamma-ray regions.) Second, with the commercial exploitation of various substances such as silicon and other more exotic materials by the electronics industry, we have developed sensors capable of detecting radiation at all wavelengths in the spectrum. By combining these electronic sensors with large ground- and space-based telescopes, astronomers ushered in a revolution in our understanding of the cosmos.

This technological revolution has the potential to do for astronomy and our understanding of the Universe what the Earth-circling voyages of the late Renaissance did for geography and our knowledge of the Earth. The voyages of the great explorers such as Columbus and Magellan permitted humans to explore and map the globe. Before these voyages, our knowledge of the Earth and its geography was limited to small regions of the planet. By the end of the great explorations that started 500 years ago, we had discovered all of the major continents, islands, oceans, and seas. By the end of the twentieth century, we charted every corner of the planet. Geography evolved from exploration to consulting maps.

The revolution in astronomy made possible by our new-gained ability to see all wavelengths of the spectrum promises a similar profound transformation of our understanding of the cosmos. Today's giant telescopes have the potential to observe most of the volume of the Universe. With giant silicon cameras, we will be able to chart the location and properties of most of the ordinary matter in the Universe.

The first crude maps of the cosmic microwave background already exist, and better maps are being made by satellites such as WMAP (the *Wilkinson Microwave Anisotropy Probe*). As the National Aeronautics and Space Administration (NASA), the European Space Agency (ESA), and the space agencies of other countries launch new space telescopes, and as ever more powerful ground-based facilities are built, detailed information on the evolution from the distant and ancient Universe to the present will be obtained. Unlike the geological record of our own planet's history, which is difficult to read and interpret correctly, telescopes enable the direct observation of the evolution of matter from 300 000 years after the Big Bang to the present. The beginning of the third millennium promises to be the golden age of astronomy.

The process of scientific discovery that started at the end of the Middle Ages is continuing. The motions of objects on Earth, and the celestial bodies in our solar system, motivated Galileo and Newton to develop mechanics and the calculus. As our understanding of Nature evolved, so did technology. With the Industrial Revolution came the steam engine and the exploitation of the powers of gases and thermodynamics. During the nineteenth century, James Clerk Maxwell united the budding understanding of electrical and magnetic phenomena into his electromagnetic theory, forming the theoretical underpinnings for the utilization of electromagnetic forces for the betterment of life. Electric lights, the telephone, radio, television, and our telecommunications and broadcast industries were the commercial end results. As the nineteenth century turned into the twentieth, Einstein, Bohr and other scientists laid the foundations of "quantum mechanics" that describe atoms. The science of the atom in turn revolutionized chemistry and led to inventions such as the transistor, and to today's amazing developments in electronics. The electronics industry is the source of the multi-wavelength sensors that have proven crucial for astronomy. From the wellspring of technology, we derive the tools that enable us to remotely sense the Universe.

Telescopes

The current revolution in astronomy is driven by technology. The key ingredients are the development of new light sensing devices, the availability of cheap and fast computers, the increasing light collecting areas of new telescopes, access to space, and our growing ability to sense all wavelengths of the spectrum. New technology creates unprecedented opportunities for advances in modern astronomy. Our current understanding of astronomical processes is the culmination of four centuries of technical and scientific advance. The first great technological leap was the development of the telescope in the early seventeenth century. Though very crude by modern standards, Galileo's small telescope was powerful enough to reveal a host of new phenomena that had never before been seen by human eyes. The first telescopes utilized ground and polished lenses developed to correct poor eyesight. Thus, the telescope represents an instance where the modification of a

Figure 2.1. The old 6 inch refractor at the Mt. Wilson Observatory in California. (Jack Eastman).

practical device led to the invention of a powerful new scientific tool. The type of telescope used by Galileo is called a "refractor" since its operation depends on the bending (the *refraction*) of light by curved lenses made from glass. Refracting telescopes (Figure 2.1) utilize a large lens at the front of the instrument, an *objective*, to concentrate and focus the incoming light. For the human eye to see an image, a second lens (or group of lenses), called an

Figure 2.3. The dome of the Kitt Peak 4 meter reflecting telescope near Tucson, Arizona. (NOAO/AURA/NSF).

of wavelengths (or colors) than either the human eye or photographic film. Thus, these image sensors are nearly ideal for the recording of faint astronomical objects at wavelengths in the 0.3 to 1 micron range.[5]

Opening the spectrum

Until 1929, all astronomy utilized light with wavelengths near those sensed by the human eye. In that year, the first radio signals were recorded from

Figure 2.4. The 4 meter Mayall reflecting telescope at Kitt Peak showing the Cassegrain cage located behind the primary mirror. Many of the images in this book were obtained with this telescope using the prime focus CCD camera called MOSAIC. (NOAO/AURA/NSF).

natural sources beyond the solar system, giving rise to a new way of looking at the sky: radio astronomy. After World War II, this new discipline developed rapidly. Radio telescopes use large dishes which act like the mirrors in an ordinary reflecting telescope (Figure 2.6). However, the lengths of radio waves are measured in centimeters or meters instead of microns. Thus, the reflecting surfaces of radio telescopes can be considerably rougher and less accurate than those of visual wavelength telescopes. As a result, radio telescopes can be very large, with diameters reaching about 100 meters. Even so, at long radio wavelengths, such giant dishes provide resolutions considerably worse than that of the human eye.[6] Indeed, the early views of the radio sky were very fuzzy when compared to visual wavelength images.

Figure 2.5. The summit of Mauna Kea showing the observatory domes at night. (Wei-Hao Wang).

Figure 2.6. The 100 meter diameter Green Bank Radio Telescope located in Green Bank, West Virginia. (NRAO/AUI).

Soon after World War II, radio astronomers developed a technique for electronically linking radio telescopes separated by many miles. The resulting instrument, called a radio *interferometer*, can measure the locations of radio sources with nearly arbitrary precision ranging from a few arcseconds in the 1950s to nearly a *micro* arcsecond today.[7] For the first time, radio sources could be identified with objects also seen on visual wavelength images. These

Figure 2.7. The Very Large Array radio telescope located near Socorro, New Mexico. (NRAO/AUI).

observations indicated early on that most radio sources were not just common stars. Rather, they tended to be exotic objects such as the Crab nebula, the remnant of a massive star that exploded in the constellation of Taurus in the year 1054, or distant colliding galaxies, or galaxies containing supermassive black holes[8] in their centers.

By linking many radio telescopes together in an interferometer, radio astronomers essentially mimic the performance of a dish that is the size of the separation of the different telescopes, which is sometimes tens of kilometers (Figure 2.7).[9] Such instruments enable astronomers to obtain pictures of the radio sky that are as sharp as those produced by the largest and best ground-based telescopes. The *Very Long Baseline Array* links 10 radio telescopes to synthesize an aperture that is nearly as large as the Earth. With a dish in Hawaii, another on St. Croix in the Caribbean Sea, and eight others in the continental USA, this instrument can produce pictures that resolve details a hundred times smaller than the sharpest pictures obtained with the Hubble Space Telescope.

Our recently gained ability to detect virtually all wavelengths of the electromagnetic spectrum is one of the pillars on which the on-going revolution in astronomy rests. Another is the development and application of computers that enable detailed calculations and data processing. Computers have also altered the way we think about telescopes, instruments, and spacecraft. Instead of massive structures and thick mirrors that are intrinsically rigid, modern telescopes often use lightweight and flexible frames and thin mirrors which are actively compensated for deformation by gravity under continuous computer control.

By the end of the nineteenth century, refracting telescopes reached their maximum apertures. Lenses larger than about a meter turned out to sag too much under their own weight to produce satisfactory images. Mirrors, on the other hand, could be supported from the back, and the diameters of reflecting telescopes reached 5 meters by the middle of the twentieth century. However, their cumbersome mounts in which one axis pointed towards the celestial pole (called *equatorial mounts*),[10] and the enormous mass of their optics, appeared to make larger instruments impractical.

Radio astronomers, by working at longer wavelengths, could tolerate much larger deflections in the mounts and optics of their telescopes. In addition to building giant dishes, they also started using altitude–azimuth (alt–az) mounts in the 1960s. Though these telescopes' supports required motion about two axes to track stars across the sky (instead of the single axis motion of an equatorial mount), they were much cheaper and easier to build, and enabled radio telescopes with diameters as large as 300 meters to be constructed.[11] A key innovation of radio astronomers was the use of individual small curved panels to tile the reflecting surface of radio telescopes. Thus, giant reflectors could be assembled on site from smaller pre-fabricated parts.

By the late 1980s, some visual wavelength telescopes also adopted alt–az mounts and the segmented mirror approach. The world's largest reflectors (such as the two 10 meter diameter Keck telescopes) use dozens of hexagonal mirror segments that are brought into precise alignment with computer control. Segmented optics, lightweight support frames, and computer controlled alignment and flexure control, have created unprecedented opportunities to build even larger telescopes than the 8 to 10 meter diameter visual wavelength instruments that have been commissioned within the last decade.

Already under study are visual and near-infrared wavelength fully steerable instruments with segmented mirrors 30 to 100 meters (100 to 300 feet) in diameter. These telescopes will have primary mirrors as large as football fields, and yet their entire surfaces will be held to within a millionth of an inch of a prescribed figure. The largest such concept is the European Southern Observatory's *OverWhelmingly Large* (OWL) telescope design.

To utilize such giant apertures fully, methods are being developed to compensate for and remove the blurring effects of turbulence in the Earth's atmosphere. Already some ground–based images, corrected by such "adaptive" optics,[12] in some ways rival the very sharpest pictures returned from the Hubble Space Telescope that flies above the turbulent air.

Space

The 1960s saw the dawn of the Space Age. With the advent of rockets and satellites, astronomers gained access to entirely new portions of the spectrum which are blocked by the Earth's atmosphere. The first X-ray, gamma-ray, and infrared sensors were flown. For the first time, astronomers could study the entire spectrum of electromagnetic energy that reaches the Earth from the cosmos. The birth of space astronomy heralded a decade of exploration during which many new astronomical phenomena were discovered. During subsequent decades, a steady stream of astronomical missions would continue the exploration of these new parts of the spectrum and the phenomena that were discovered.

The *Uhuru* X-ray satellite, launched in 1970, mapped the X-ray sky and uncovered hundreds of remnants of massive stars that had exploded, the first black hole candidates, and distant galaxies containing black holes. A successor mission, NASA's *Einstein* X-ray observatory discovered that young low-mass stars produce powerful flares of X-rays. Today, X-ray astronomers are using the *Chandra* X-ray Observatory to obtain the deepest and highest resolution X-ray images of the sky ever. The European Space Agency's X-ray Multi-Mirror Mission, called *XMM-Newton*, is providing the most sensitive spectra of X-ray sources.

At even shorter wavelengths, access to space has enabled the investigation of the most energetic photons that exist: gamma rays. The *Vela* satellites, flown by the US Department of Defense to monitor violations of the nuclear test ban treaties, discovered flashes of intense gamma-ray emission that came to be known as gamma-ray bursts. Subsequent studies with more advanced satellites such as NASA's *Compton Gamma Ray Observatory* provided new insights into this exotic phenomenon and as a result, within the last few years, we started to better understand gamma-ray bursts. There is growing evidence that these energetic flashes are produced by the death throes of the most massive (and therefore youngest) stars in very distant galaxies. For a few seconds, gamma-ray bursts may release more energy than the combined light of all the stars in the Universe.

Space has also enabled phenomenal advances in infrared astronomy. At wavelengths longer than visible light, the *Infra-Red Astronomy Satellite* (IRAS) obtained during the 1980s the first survey of the entire sky at infrared wavelengths 25, 50, 120, and 200 times the wavelength of visible light. For the first time, the intense glow of interstellar dust in our Milky Way was revealed with a resolution comparable to that of the unaided human eye. IRAS detected

Figure 2.8. A photograph showing the Hubble Space Telescope used to obtain many of the images shown in this book. (NASA/STScI).

thousands of objects which we now know are stars and planetary systems in the process of birth, condensing from the giant clouds first detected by radio telescopes sensitive to the millimeter wavelength emissions of simple molecules. IRAS was followed by several increasingly sophisticated infrared space telescopes. These first and second generation infrared space telescopes have mirrors less than a meter in diameter.

The crown jewel of NASA's current space science program is the *Hubble Space Telescope* (HST – Figure 2.8). Conceived during the 1960s soon after the start of the Space Age, HST was finally launched by the Space Shuttle in 1990. By flying above the atmosphere, Hubble's images are clearer than even those produced by the largest ground-based telescopes, which have to contend with turbulence in the Earth's atmosphere, even though the HST mirror is only 2.4 meters in diameter.

The various space agencies are presently investigating novel optical technologies that may enable much larger mirrors to be built and deployed in space. On the drawing boards is the *James Webb Space Telescope* (JWST) which will use a segmented mirror built up of smaller self-deployed optical elements. This infrared optimized facility, which will be launched early in the next decade, will have a mirror a bit larger than 6 meters in diameter and will orbit the Sun at a distance of about a million kilometers from us (about three times the Earth–Moon separation) so that heat radiation from the

Earth and Moon will not affect it significantly. Even larger space telescopes that use hair-thin "gossamer" mirrors are under consideration.

The twenty-first century is likely to see remarkable developments in astronomical instrumentation. By linking large space telescopes into giant interferometric arrays, we may eventually "synthesize" apertures many kilometers or even thousands of kilometers in diameter.

Today, we measure resolution in arcseconds. The Hubble Space Telescope and ground-based facilities assisted by adaptive optics[12] can deliver images with about 0.1 arcsecond resolution. Radio interferometry is now pushing into the *micro*-arcsecond domain, and some pictures of star forming regions and jets produced by giant black holes located in the centers of some galaxies (see Chapter 15) have resolutions about ten thousand times better than those produced by the Hubble Space Telescope. During the next decades, giant interferometers in space operating at X-ray, visual, and near infrared wavelengths may push resolution into the regime of *one-billionth* of an arcsecond! We will be able to take pictures that resolve fine details on the surfaces of other stars, obtain images of distant planets orbiting other stars that are as sharp as the pictures of Earth returned by our meteorological satellites, and observe the inner disks spinning around black holes. We will use these telescopes to peer into the past to piece together the detailed history of our Universe from the birth of the very first stars to the present.

Instruments and computers

Novel detectors are also under development. As amazing as CCDs are, there are hints of even more powerful detector technologies. Physicists are investigating devices that not only can use all of the incident light to record an image, but can also determine both the exact arrival time of each photon and measure its energy (or wavelength). When such sensors become available, we will be able to obtain the spectrum of every object in an image. Instead of a picture, these sensors will produce *data cubes* that will require virtual-reality techniques to display.

The use of these detectors on the next generation of 10 to 30 meter class telescopes will enable astronomers to map the Universe in ways we can barely comprehend now. Every galaxy in the Universe intrinsically brighter than about 1 to 10 percent of the Milky Way can be imaged. Their locations and recession speeds, their composition, and other characteristics will be charted. The resulting "Map of the Universe" may have an impact similar to the charts that resulted from the great voyages of discovery half a millennium ago. These globe circling journeys enabled us to map the Earth. In a similar fashion, the next generation of astronomers have the opportunity to chart the Universe.[13]

There are four distinct styles of astronomical research in the quest to understand the phenomena of the Universe. While *observers* use telescopes to gather data and *instrumentalists* build telescopes, cameras, and spectrometers used by the observers, *theorists* explain or predict the phenomena seen by the observers, and *numericists* use computers to construct models of these phenomena.

Just as electronics has revolutionized instruments and observation, the computer has also brought about its own revolution in the interpretation of observations. With the aid of computers, astronomers can make detailed predictions of the future motions of planets, moons, artificial satellites, asteroids and comets in our solar system. They can model the thermonuclear fusion processes that occur within normal stars, and follow the evolution of these objects as they consume their thermonuclear fuel. They can explore stellar death and re-create nova and supernova explosions in computer simulations which show details far beyond our ability to observe. Models of the early Universe simulate the formation of galaxies. Such models can be compared to observations to pin down the correct theory of galaxy formation and evolution. The numericists can even model processes that are hidden from us by the fog of the Cosmic Microwave Background. They are now applying the tools of their trade to the simulation of the birth of stars and planets.

Computers, the availability of relatively inexpensive CCD cameras, and modern photographic materials have enabled the acquisition of spectacular images of astronomical objects with relatively modest equipment and small telescopes. Indeed, many of the images shown in this book were acquired with small telescopes or lenses. Not only can such equipment be used to obtain stunning images, but it can also produce first-rate research results. Observing time on large telescopes is an expensive commodity. As a result, such facilities are not well suited for long-term brightness monitoring programs. On the other hand, small telescopes equipped with modern CCD cameras can also provide precision measurements of the light output and variations of stars. Such equipment is ideal for measuring the *light curves* of variable stars, or for the detection of the sudden appearance of a new nebula, a nova, or supernova. With patience and care, extra-solar planets can be detected by means of their transits across distant stars (see chapter 13). Despite their diminutive size, small telescopes have once again become powerful scientific tools when used with understanding and perseverance.

Astronomy in the twenty-first century promises to be more exciting than ever.

3 The dark clouds of the Milky Way

On a clear moonless night, at a dark place far from city lights, we can see the Milky Way as our ancestors must have seen it (see Figure 3.1). The Milky Way appears as a bright band of diffuse light stretching across the sky. What we see is the combined light of billions of distant stars located within our Galaxy. When we look closer, we see that the shining band of the Milky Way is bifurcated by dark bands and patches. At the beginning of the twentieth century there was a heated debate whether these vacancies between the stars were caused by an actual absence of stars, or were produced by intervening material that obscures the glowing background of stars. We now know that these dark areas contain gigantic clouds of gas and dust interspersed between the stars. It is out of these dark clouds that stars are born.

Dark clouds and the structure of our Milky Way galaxy

Dark clouds are among the coldest objects in the Universe. Their temperatures are typically only 10 degrees or so above absolute zero, the temperature at which atoms come to a standstill. The most common constituent of dark clouds is hydrogen, which is also the most abundant element in the Universe. At the very low temperatures of dark clouds, most hydrogen atoms become bound to each other in pairs, forming hydrogen molecules. Dark clouds are mostly made of molecules of hydrogen, with a sprinkling of molecules of many other far less common elements. Therefore, dark clouds are mostly *molecular clouds*.

At very low temperatures, molecular hydrogen does not emit or absorb light at visible wavelengths. In fact, cold molecular hydrogen is hard to detect at any wavelength. Fortunately, there are two ways that Nature has helped astronomers in their studies of molecular clouds. First, dark clouds are surrounded by envelopes of hydrogen atoms which are not bound up in molecules, and *single* hydrogen atoms emit copious radiation at centimeter wavelengths. Second, dark clouds are spiced with small amounts of carbon monoxide (CO), allowing astronomers to peer into their interiors, because this molecule emits bright radiation at millimeter wavelengths. CO has become the most important tracer of the structure and kinematics of molecular clouds. Additionally, more than 100 other molecules have been discovered in small quantities inside dark clouds.

Figure 3.1. The glorious summer Milky Way passes through the constellations Ophiuchus and Scorpius and close to the bright red star Antares. (NOAO/AURA/NSF).

Radio astronomy developed rapidly after World War II, as discussed in Chapter 2. The 21 centimeter wavelength emission of atomic hydrogen was discovered along the Milky Way immediately following the war. As radio telescopes became more sensitive and worked at shorter wavelengths, astronomers discovered radio emission from molecules in interstellar space in the 1960s. By 1970, they found that dark clouds produce copious emission from the bright 2.6 millimeter line of carbon monoxide,[1] and soon they discovered that molecular clouds can attain gigantic proportions. It then became clear that the Milky Way contains several billion solar masses of cold molecular hydrogen that could be traced by CO observations.

Radio techniques are now so refined that astronomers routinely map molecular clouds and their atomic envelopes. They can determine their internal structure and estimate how massive such clouds are. But, perhaps most importantly, they can also measure the speed of emitting gas relative to us. The principle is exactly the same as the one behind the changing pitch of a

Figure 3.2. The spiral galaxy M83. Note the dark clouds and bright stars in the spiral arms and the central bulge of older, yellow stars. (ESO).

moving siren. When an ambulance is moving towards us, the tone of its siren becomes more high-pitched. Conversely, when the siren moves away the pitch gets lower. Similarly, the radio waves coming from dark clouds are slightly shorter or longer if the clouds move either towards or away from us. This Doppler effect[2] enables the measurement of motion along our line of sight.

By meticulously measuring the radio signals from dark clouds along the Milky Way, astronomers have assembled a map of their locations and motions. The clouds move around the center of our Milky Way and are concentrated in spiral arms. We have learned that the Milky Way is a *spiral* galaxy. Figure 3.2 shows a nearby spiral galaxy which resembles our Galaxy, with bands of dark clouds and luminous massive stars stretched out along the spiral arms. Our Sun, located roughly half way between the center and outer edge of our Galaxy, completes one journey around the Galactic Center every 250 million years. With a diameter of at least 100 000 light years, the Milky Way is enormous. But it is very thin, with the molecular clouds confined to a layer only about 300 light years thick.

Molecular clouds

Molecular clouds appear dark when seen against the rich stellar tapestry of the Milky Way because they contain minute amounts of dust mixed with all the gas. These solid particles originally condensed in the atmospheres of cool giant stars and were blown into the realm between the stars by stellar winds.[3] Most grains are made of graphites and silicates and are much smaller than usual household dust; interstellar grains have sizes less than one-ten thousandth of a millimeter. Although only about 1 percent of the mass of a cloud consists of dust, this tiny quantity is enough to absorb a lot of star light. Figure 3.3 shows a photograph of a region along the Milky Way where a dark cloud blots out the light from numerous distant stars. Where the cloud is thick and contains much dust along our line of sight, we see no background stars. Where only a thin layer of dust intervenes, the cloud is translucent and some background stars can be discerned.

This obscuring effect of dust in molecular clouds has provided another way to study their structure. One can simply count stars seen through the cloud and compare the numbers to counts in an unobscured nearby region. Detailed cloud maps can be made this way. However, where the cloud is totally opaque, no information about its structure can be extracted. Finally, the star counting method assumes that all the stars seen in the direction of the cloud are *behind* the cloud. This is true only for the nearest clouds. Figure 3.4 shows a pair of nearby clouds, whose outer layers are sufficiently thin that the light from the numerous background stars can pass through. It is evident that the light from these background stars is reddish. This is because dust gradually becomes more and more transparent with longer wavelengths of light. In other words, red light can shine through a cloud more effectively than blue light, making stars seen through a cloud appear redder.

Dust can also be observed directly, not only through its obscuring effect. Because dark clouds are very cold, the dust grains are also very cold. All bodies emit thermal radiation, but the colder they are, the longer is the wavelength of the emitted light.[4] Thus, by measuring dark clouds far into the infrared and submillimeter wavelength regimes, we can directly detect the presence of cold dust in dark clouds.

Infrared and radio wavelength studies provide the best knowledge of the properties of the clouds of gas and dust in interstellar space. Figure 3.5 shows an infrared image of the entire sky projected onto the paper with the plane of the Milky Way defining a horizontal line along the major axis of the figure. In this projection, the center of our Galaxy lies in the middle of the figure, the north pole of the Galaxy lies at the top, and the south pole lies at the bottom. The colors correspond to the wavelength; blue is near 10 microns while red is near 100 microns. The yellow band in the middle shows the cool dust clouds of the Milky Way. In the figure, interplanetary

Figure 3.3. A dark cloud, Barnard 86, is seen against the dense tapestry of distant stars in the Milky Way. A small cluster of stars, NGC 6520, is seen nearby. (J.–C. Cuillandre/CFHT).

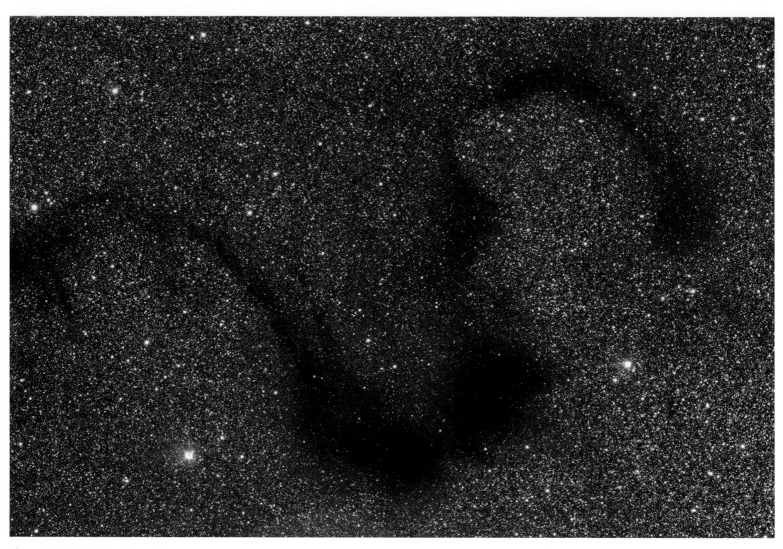

Figure 3.4. A dark cloud, Barnard 72, seen in silhouette against the Milky Way. (J.-C. Cuillandre/CFHT).

dust emission has been removed to show the galactic dust emission more clearly.

Figure 3.6 shows the sky at a wavelength of 21 cm where atomic hydrogen produces strong emission. This figure depicts the distribution of several billion solar masses of atomic hydrogen in our Galaxy. Like the dust emission, this species is also concentrated in the Galactic plane. Similarly, Figure 3.7 shows the distribution of CO molecules.

Comparison of Figures 3.5 through 3.7 demonstrates that the molecular gas traced by CO emission is more confined to the thin plane of our Galaxy than either atomic hydrogen, dust, or stars. Hydrogen can be seen towards every direction. As will be discussed in Chapter 14, atoms and molecules cycle between these various phases of the interstellar medium. Initially, molecular clouds condense from atomic hydrogen gas. But as stars form, the molecular clouds are destroyed and their atoms are eventually recycled back into the atomic hydrogen phase.

Combining studies of dark clouds using radio telescopes with maps from star counts has given astronomers a detailed look at their structure and

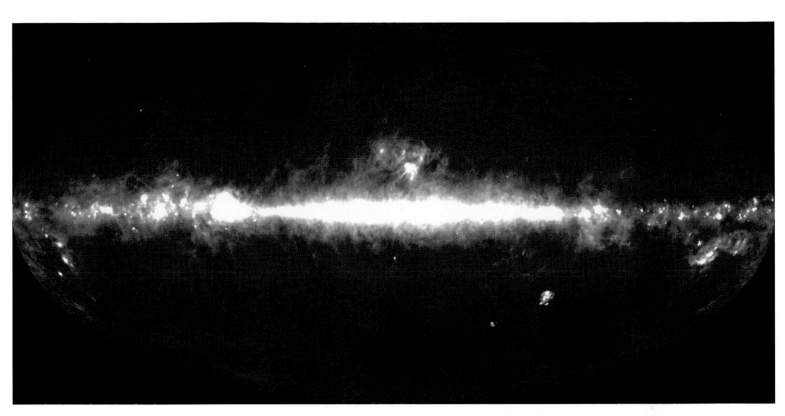

Figure 3.5. An all-sky image of infrared dust emission obtained with the COBE satellite. The plane of the Milky Way lies along the major axis of the projection. The Galactic center lies in the middle and the Orion clouds are located just below the right edge of the figure. (NASA/COBE).

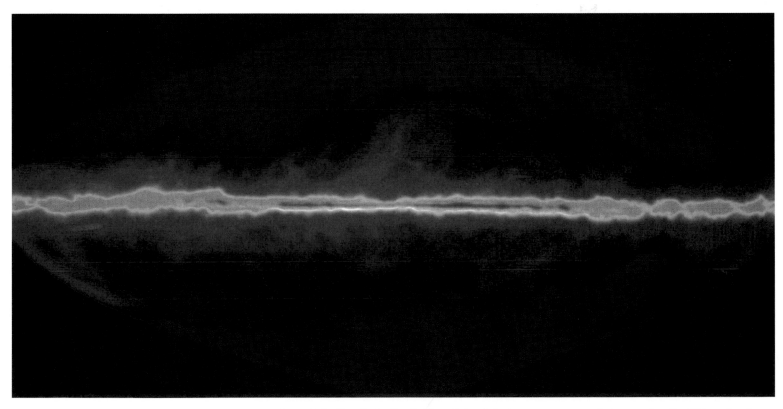

Figure 3.6. The all-sky distribution of atomic hydrogen as measured at a wavelength of 21 centimeters with a radio telescope. (NRAO/NSF).

35   The dark clouds of the Milky Way

Figure 3.7. The all-sky distribution of molecular clouds as traced in the 2.6 millimeter wavelength emission of the carbon monoxide molecule, a robust tracer of molecular hydrogen. (T. Dame).

properties. Clouds come in a variety of shapes and sizes. The largest are called *Giant Molecular Clouds* and are the largest objects in our Galaxy. Giant molecular clouds have extents of up to 300 light years and may contain over a million times the mass of our Sun. The shapes of these clouds are not easy to describe, because they do not have well defined edges. Their structure is complex, with dense concentrations connected by sheets and filaments of gas and dust. This chaotic structure is seen on all scales. Some astronomers have compared molecular cloud complexes to fractal figures in which intricate structures continuously repeat themselves on smaller and smaller scales. Figure 3.8 shows a millimeter wavelength image of a pair of giant molecular clouds in the constellation Orion.

On the other end of the size scale, one finds the very small *Bok globules*, named after the Dutch astronomer Bart Bok, who studied them. These tiny clouds are no more than a light year across (size is of course a relative concept, and while small compared to giant molecular clouds, globules are still thousands of times the size of our Solar System). Globules can be compact and enormously dense. When the eighteenth century British astronomer William Herschel first encountered one while looking through his telescope, he exclaimed, "But, there is a hole in the heavens!" (see Figure 3.9). Bok globules typically have masses of only a few Suns.

In between the giant molecular clouds and the small Bok globules, there is a large range of cloud masses. Isolated dark clouds with masses of 10–100 solar masses are common, as are aggregates of dark clouds with masses from a few hundred to several thousand solar masses. Although the smaller clouds are much more common than the giant molecular clouds, these latter structures are so enormous that the bulk of the molecular gas in our Galaxy is found in giant molecular clouds.

Like the Earth, the Sun, and all other celestial bodies, molecular clouds have gravitational fields. With masses many thousand times larger than the Sun, gravity plays a fundamental role in how molecular clouds evolve. Why don't clouds collapse under their own weight? With the low temperatures of dark clouds, the pressure associated with random motions of molecules[5]

Figure 3.8. A 2.6 millimeter wavelength image showing most of the constellation of Orion in the light of carbon monoxide. (J. Bally).

is also low and insufficient to support a cloud. We would therefore expect that clouds should rapidly and efficiently collapse under their own weight and turn into many thousands of stars. But it turns out that there are several other factors which stabilize molecular clouds against gravitational collapse. One is *turbulence*, a constant churning of the gas and dust in molecular clouds, which pushes the material in clouds in chaotic patterns. Turbulence

Figure 3.9. The dark cloud Barnard 68 as observed with the European Southern Observatory's Very Large Telescope. (J. Alves/ESO/VLT).

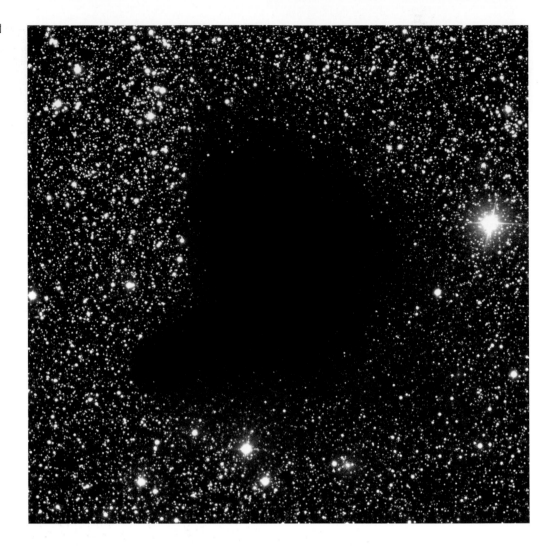

is still somewhat mysterious. Although it can easily be measured, we do not fully understand how it originates and how it is maintained. Possibilities include outflows from young stars (see Chapter 6), radiation fields, explosions, stellar winds, and gravitational collapse. Energy to maintain turbulence can be injected from inside the cloud or from the outside, on small scales or large. Though the sources are not understood, it is clear that turbulent pressure helps to support clouds against their own gravity.

Magnetic fields provide another important supporting agent in clouds. Interstellar clouds are threaded by magnetic field lines. Although these fields are weak, their effect can be significant. A small fraction of the gas in a molecular cloud is ionized, that is the atoms have lost one or more of their electrons. These ions and electrons couple to the magnetic fields, so that they are forced to move along the field lines. Magnetic field lines resist getting close to each other, thus creating a magnetic force that inhibits compression of the gas. It is not easy to measure magnetic fields in dark clouds, and only in recent years have good measurements been made. Although magnetic fields are invisible, it is tempting to speculate that some filamentary structures in dark clouds may result from magnetism (see Figure 3.10).

Figure 3.10. A portion of the North America / Pelican Nebula in Cygnus as seen in the red emission of hydrogen. Note the linear filaments of dust. Magnetic fields may be responsible for organizing the foreground dust cloud. (J. Bally & B. Reipurth, NOAO/AURA/NSF).

Cores, dust, and chemistry

Observations of dark clouds have revealed that they are highly complex structures. On the largest scales, molecular clouds show many clumps of varying sizes, often connected by filaments and surrounded by more tenuous material. These structures partly relate to how the clouds formed (see chapter 14), but also reflect their subsequent evolution and the effect of star formation. The clumps are often so large and dense that light from background stars cannot pass through them, and so in visible light they appear uniformly dark. But detailed observations of molecules at millimeter wavelengths have shown that the interior of clumps themselves can be highly structured, often containing one or more very dense cores. It is out of such cores that stars are born.

Molecular clouds show fragmentation with a hierarchy of sub-structures. Determining how clouds gain their fragmented structure casts light on how stars form. There are two basic mechanisms that can fragment a cloud. The first is driven by gravity itself. If a molecular cloud after its condensation from the interstellar medium has small fluctuations in density, then further contraction by gravity will amplify these variations, leading to an increasingly fragmented structure. The second mechanism to form structures in clouds may be driven by the turbulent motions mentioned in the previous section. If the gas in a molecular cloud is churning with velocities that exceed the sound speed,[6] then powerful shock waves[7] will form. Such shock waves will compress the gas and a hierarchy of compressed regions or sub-structures can result. Almost certainly both these mechanisms play a role in creating the structured appearance of molecular clouds, ultimately leading to the birth of stars.

The outer layers of dark clouds are exposed to a hostile environment. Powerful ultraviolet radiation, X-rays, and energetic high-energy particles[8] can break up, or dissociate, molecules, and molecular clouds are therefore surrounded by large tenuous envelopes of atomic hydrogen, with densities of only a few atoms per cubic centimeter or less. But the gas density dramatically increases in the clumps and even more in the small cores, where it can exceed 100 000 molecules per cubic centimeter. Such different physical environments also lead to differences in the chemistry of the gas. As the gas density increases, molecules collide with dust grains with greater frequency, and begin to stick to the grains. This has two effects. First, the molecules that stick to grains are removed from the gas. And second, the molecules stuck on grains can interact, forming new molecular species. Some of these may occasionally be dislodged from the grain surfaces, as grains are heated by incident radiation and high energy particles. As the structure of a cloud evolves and cores form, a complex chemistry develops and the composition of the gas in the cloud becomes transformed. Simple molecules such as CO are depleted and more complex species such as formaldehyde, methyl-cyanide, and methanol appear. Changes in chemistry become pronounced

Figure 3.11. The Ophiuchus molecular cloud complex is a small, but nearby dark cloud, which is forming a small cluster of about one hundred stars. (Wei-Hao Wang).

after a young star is born and starts to heat the surrounding dust with its radiation. As a result, grains release a wealth of exotic molecules that play important roles in the evolution of young stars and their future planetary systems.

Dark clouds: cradles of stars

Young stars are usually found in or near molecular clouds. Thus, the formation of stars is related to processes occurring in these objects. Although there are many more smaller clouds, most of the Milky Way's molecules are found in giant molecular clouds. The vast majority of young stars are born in giant molecular clouds which are therefore the dominant cradles of stars. The Orion star forming region, located about 1500 light years from the Sun, is the closest giant molecular cloud complex. As we shall see in later chapters, this complex has produced many thousands of stars in the last 10 million years. But because small clouds are more abundant, the nearest young stars are found in the smaller cloud complexes such as those in Taurus and Ophiuchus that are located only 300 to 400 light years away. Figure 3.11 shows the nearby Ophiuchus cloud complex, which is actively forming low-mass stars. Though these closer clouds do not produce nearly as many stars as the giant clouds in Orion (each cloud has produced only about a hundred stars) their proximity makes them excellent targets for study. Much of what we know about the birth of low-mass stars has been learned from these investigations. Also, several small groups of stars with ages of only tens of million years have been found a mere 150 light years from us, but the dark clouds from which these very nearby young stars originally formed have been completely dissipated by now.

Only very few giant molecular clouds are known that are not forming stars. It follows that star formation starts in a cloud almost immediately after its formation. The cloud will then produce stars for several million years, but typically dissipates in less than about 10 million years, since groups or clusters of young stars never have ages that span more than 10 million years. As we shall see in coming chapters, young stars destroy their habitats, especially when massive stars form. The flood of ultraviolet radiation they release can sculpt the surrounding cloud into gracious, windblown structures (Figure 3.12) that soon disappear. In the process, the birth of new generations of stars can be triggered in a vast cycle, as discussed in detail in chapter 14. Cloud cores close to newborn massive stars can be explosively torn apart by the ferocious onslaught of the powerful radiation fields (Figure 3.13). Thus, clouds are transient structures. They are born out of the surrounding interstellar medium, spend most of their short lives producing stars, and eventually dissipate back into the interstellar medium.

How do giant molecular clouds turn into stars? At what rate does star formation occur? How much of a cloud's mass is converted into stars before it is destroyed? What determines the masses of stars? These questions are the subject of intense and ongoing research activity. Although we have solid ideas about the star formation process, many important details are only poorly understood. As discussed in the next chapters, stars tend to form in groups. But to avoid the enormous complexity of clustered star formation, most observational studies and theoretical modeling have focused on single, isolated star formation in dark clouds. While this provides an incomplete picture, it is a tractable approach that provides much insight into the star formation process. We will discuss the birth of single, isolated stars in the next chapter.

Figure 3.12. A large cometary globule in the HII region IC 1396 is being shaped by powerful ultraviolet radiation from nearby massive stars outside the picture. (J.-C. Cuillandre/CFHT).

Figure 3.13. Dark clouds in the HII region NGC 7000 are being destroyed by their hostile environment. (J. C. Cuillandre/CFHT).

4 Infant stars

Stars form in dusty molecular cloud cores. Newborn stars are therefore deeply embedded in their parental cloud, surrounded by a cocoon of placental material. The gas and dust shrouding newborn stars tends to be so dense that any visual wavelength light is completely absorbed. However, infrared light and radio waves can, with their longer wavelengths, propagate through dusty interstellar gas with relative ease. Observers have therefore turned to these longer wavelength regimes in their efforts to discover the youngest stars still embedded in the opaque interiors of molecular clouds. Recent advances in infrared and radio technologies have revolutionized our understanding of the birth of stars.

The cold Universe: far-infrared and sub-millimeter observations

The sky looks different when viewed at visual and at infrared wavelengths. Our eyes are most sensitive at visual wavelengths because that is where the Sun emits most of its light. Therefore, our visual senses are tuned to see stars. Stars have temperatures ranging from a few thousand to several times ten thousand degrees. Cooler objects emit at longer wavelengths in the near- and far-infrared portion of the spectrum. Some creatures, such as snakes, have eyes sensitive to the heat radiation emitted by warm blooded animals. If they were to peer towards the sky, they would see a scene quite different from what we see. While at visual wavelengths the appearance of the sky is dominated by stars, at infrared wavelengths beyond 10 microns the sky is filled with the radiation of giant clouds of interstellar dust and a few very cool stars.

The exploration of the sky in the infrared portion of the spectrum began in the 1960s. Until the early 1980s, all infrared observations were obtained with single pixel detectors. But in the last few decades, infrared light sensors with an ever increasing number of pixels have become available. We can now routinely obtain digital images at near-infrared wavelengths between 1 and 5 microns, just beyond the visible portion of the spectrum.

In their quest to detect ever cooler objects, astronomers have sought sensors operating at longer wavelengths. Though most infrared sensors were initially developed for the military, they eventually became available to

Figure 4.1. The Gemini North telescope is optimized for infrared observations and is located at the very high and very dry Mauna Kea volcano in Hawaii. (Gemini/AURA).

astronomers, enabling them to push observations into the mid-infrared (about 5 to 30 microns) and far-infrared (about 30 to 300 microns) spectral domains. But here one encounters a fundamental problem: the Earth's atmosphere blocks almost all mid- and far-infrared radiation.

At best, ground-based observations at infrared wavelengths beyond a few microns are difficult because water vapor absorbs infrared radiation. It helps to place infrared observatories at very high and dry sites; the Mauna Kea volcano in Hawaii, located at 4200 m (13 700 feet), is one of the best ground-based locations for infrared observations, and observations can be pushed to wavelengths as long as 24 microns. Figure 4.1 shows the Gemini-North Telescope, a telescope optimized for infrared observations and located on top of Mauna Kea. Even dryer sites such as the South Pole are being used for some measurements in this wavelength region. Pushing towards ever higher altitudes, NASA has built an airborne observatory, the Stratospheric Observatory For Infrared Astronomy (SOFIA) which has a 2.5 m diameter telescope mounted in the tail of a modified Boeing 747 aircraft. Balloons have carried infrared telescopes into the stratosphere, reaching altitudes of more than 100 000 feet.

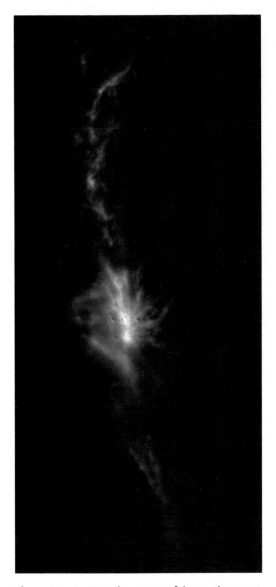

Figure 4.2. An 850 micron map of the northern part of the Orion A molecular cloud. The Orion Nebula lies in front of the brightest dust emission located in the middle of this image. (D. Johnstone and J. Bally).

But even from these extreme environments, the conditions for infrared observations are not ideal. At near-infrared wavelengths, the upper atmosphere glows with bright emission from the hydroxyl molecule which produces the equivalent of twilight conditions for visual observers. At wavelengths longer than 3 microns, the atmosphere and telescope glow with thermal infrared emission, making observing from the ground or from airborne platforms similar to visual-wavelength observations in broad daylight through a self-luminous telescope! Only by developing special observing techniques have astronomers been able to overcome these obstacles to explore the infrared sky.

Ultimately, the ideal environment for infrared telescopes is space, beyond the absorbing effects and bright emission of the Earth's atmosphere. A number of infrared telescopes placed in orbit around Earth have been dedicated to the study of cold objects with temperatures ranging from about 20 to a 1000 degrees. These include the *Infrared Astronomical Satellite* (IRAS), the *Infrared Space Observatory* (ISO), and the *Space Infrared Telescope Facility* (Spitzer), already mentioned in Chapter 2, all of which have become household names for astronomers since they opened up the infrared window to the Universe. NASA's next big infrared telescope project, the *James Webb Space Telescope*, is a 6 m infrared telescope that holds tremendous promise for exploring the infrared sky with unprecedented sensitivity and resolution at wavelengths between 1 and about 15 microns.

Astronomers have begun to understand the various objects that inhabit the infrared sky. But they have also come to recognize that to study the very coldest objects, they must bridge the wavelength regime between the infrared and radio portions of the spectrum: the sub-millimeter region where wavelengths range from about 300 microns to 1 millimeter. Figure 4.2 shows a sub-millimeter image of the northern part of the Orion A molecular cloud (see Figure 3.8 for a 2.6 millimeter wavelength carbon monoxide image). The figure shows the location of cool dust together with small bright points where newborn stars are beginning to heat up their surroundings.

Observations at sub-millimeter wavelengths require technology that has only matured in the last decade. As in the infrared, sub-millimeter observations are adversely affected by water vapor in the atmosphere, so telescopes operating at these wavelengths must also be placed at high and dry sites. The largest ground-based astronomy project of the next decade is the *Atacama Large Millimeter Array* (ALMA), a suite of 64 12-meter antennas acting as an interferometer (Figure 4.3). Stretching across several kilometers on a high plateau in the Chilean Andes mountains, ALMA is specifically designed to work at sub-millimeter wavelengths. Its enormous size will enable ALMA to study the Universe at wavelengths down to 350 microns with a resolution that is better than that of the Hubble Space Telescope.

Armed with near-, mid-, far-infrared, and sub-millimeter techniques, astronomers are able to search the sky for the youngest stars and the clouds and cloud-cores primed for stellar birth.

Figure 4.3. An artist's view of the ALMA millimeter interferometer, a huge array of radio telescopes under construction in the Andes mountains in northern Chile. (ESO).

The search for newborn stars

Most deeply embedded, very young stars that we know today were found with the *Infrared Astronomical Satellite*. As mentioned in chapter 2, IRAS was launched in 1983 and surveyed nearly the entire sky at four mid- and far-infrared wavelengths, cataloging more than 250 000 infrared sources. Before IRAS, astronomers had only vague ideas about the appearance of the infrared sky and knew the locations of just the very brightest infrared sources. With the IRAS survey, astronomers obtained an inventory of the brighter infrared sources in the sky.

Figure 4.4 shows an optical photograph of a dense dark cloud in Cepheus. At visual wavelengths there is little indication that the cloud is in the process of actively forming stars. But IRAS located several infrared sources with signs of youth towards the cloud. The region in Perseus known as NGC 1333 has recently undergone a burst of star formation. Figure 4.5 shows an optical image of the central part of NGC 1333, where young stars illuminate their surrounding dark clouds. But at near-infrared wavelengths, a large population of newborn stars still embedded in the interior of the cloud becomes visible (Figure 4.6).

In the early 1990s, the European Space Agency launched the *Infrared Space Observatory* (ISO) to perform follow-up studies of IRAS sources and to obtain deep images and spectra. This mission provided detailed constraints on the composition of interstellar grains and ices, identified many atomic, ionic, and molecular species in cold interstellar clouds, and traced the distribution of these species with much greater precision than IRAS.

Figure 4.4. Part of the L1165 cloud complex seen at optical wavelengths. Embedded IRAS sources are marked with red dots. (B. Reipurth & J. Bally).

Another successor to IRAS was launched in 1996. A military project called the *Midcourse Space Experiment* (MSX) was designed to identify and track intercontinental ballistic missiles during their brief flight in space. To avoid confusion with cosmic sources in the background, MSX surveyed the Milky Way to characterize the astronomical targets that would confuse the identification of hostile warheads rising from the ground. Since most young stars lie along the Milky Way, this data provides another rich resource for astronomers interested in star formation.

The most ambitious infrared mission to date was launched by NASA in August 2003. The *Spitzer Space Telescope* currently obtains spectra and images

Figure 4.5. An optical image of the NGC 1333 star-forming region. (J. Bally).

Figure 4.6. An infrared image of the NGC 1333 star-forming region, showing the central area of Figure 4.5. (NASA/Spitzer).

of large portions of the Milky Way, nearby molecular clouds, and hundreds of the closest young stars (Figure 4.7). Spitzer is expected to find tens of millions of infrared sources, including galaxies, young stars, old stars, and clouds. It will probe the properties and evolution of circumstellar dust and gas around stars of all ages.

Some forming stars are so young and cool that not even these infrared space telescopes could detect them. Sub-millimeter telescopes are required to find the very youngest stars and pre-stellar cores. Only telescopes located at sufficiently high and dry places function well at these wavelengths. One is the James Clark Maxwell telescope on Mauna Kea, where conditions are so good that sub-millimeter observations are routinely performed. Figures 4.8 and 4.9 show two images of the same area near the Horsehead Nebula in Orion taken at optical and sub-millimeter wavelengths. Where the optical image shows the exterior of the cloud and a few young stars near the cloud surface, the sub-millimeter map probes the deep interior of the cloud complex and reveals numerous cool embedded sources.

Figure 4.7. An infrared image of the star-forming region RCW 49 obtained with the Spitzer Space Telescope. (NASA/Spitzer).

Protostars and spinning disks

The condensation of clouds into stars involves vast changes in size and density. The average density of hydrogen in the interstellar gas of our Milky Way is about one atom in every cubic centimeter. In contrast, the mean density of gas in stars is about twenty-three powers of ten larger. Thus, to form a star, the volume of an interstellar molecular cloud must shrink by more than twenty orders of magnitude!

Gravity is the force responsible for this enormous change in scale and density. The first stage in star birth is the formation of a molecular cloud core bound by its own gravity. Giant molecular clouds are threaded by weak magnetic fields, have chaotic structure, and turbulent internal motions, as discussed in Chapter 3. The relative motion of parcels of gas in turbulent molecular cloud interiors are typically an order of magnitude faster than the random speeds of their constituent molecules. The chaotic churning of the cloud interior implies that such parcels of gas will sooner or later collide and the strong shock waves produced by such collisions compress the gas. But the gas usually does not remain dense for long, since internal pressure and trapped magnetic fields tend to act like springs. As these shock waves pass, internal pressure acts to re-inflate the compressed layers to their previous densities. Computer-generated models of turbulent molecular clouds reveal that shock-compressed layers often form a chaotic lattice of transient sheets and filaments of dense gas. But occasionally enough matter accumulates for gravity to prevent the re-expansion, giving birth to a gravitationally bound pre-stellar cloud core. As such cores shrink and increase their density, they tend to fragment into smaller and still denser sub-cores which move about and interact.

The gravitational condensation and fragmentation of cores in a turbulent giant molecular cloud is a highly dynamic and chaotic process. As matter accumulates in some locations, so does the force of gravity, and more matter will fall in. Models show that over the course of a few hundred thousand years, collapse of dense cloud cores lead to the formation of sheets, filaments, and clumps that collide, merge, and interact in a complex interplay leading to the birth of anywhere from a few to many dozens of individual stars with a variety of masses. However, before discussing the birth of multiple stars and clusters in more detail, we focus on the fundamental processes involved in the birth of a single, isolated star from a single cloud core.

Our insights into the very earliest stages of star birth are largely based on theoretical work. This is partly due to the fact that observations probing the deepest cloud interiors are still very new, and the time-scales involved are far too long to follow observationally. But with the advent of fast computers it has become possible to simulate stellar birth from the beginning of the collapse phase.

The simplest case is the idealized situation of a spherical cloud core, where the force of gravity that acts to compress the cloud is balanced with the pressure of gas that supports the core against compression. If the core becomes denser, gravity can gain the upper hand over gas pressure, and the core begins to collapse under its own weight. On the other hand, if the temperature rises then the gas pressure also increases and collapse is resisted. The minimum mass of a cloud core that can gravitationally collapse is called the *Jean's mass*, after the British astronomer James Jean who introduced the concept in the early twentieth century. The Jean's mass depends only on the cloud density and the cloud temperature. Warm clouds with low density can

Figure 4.8. An optical image including NGC 2023 and the Horsehead. (J.-C. Cuillandre/CFHT).

Figure 4.9. The same area as in Figure 4.8 but seen at 850 micron with the JCMT in Hawaii. (D. Johnstone).

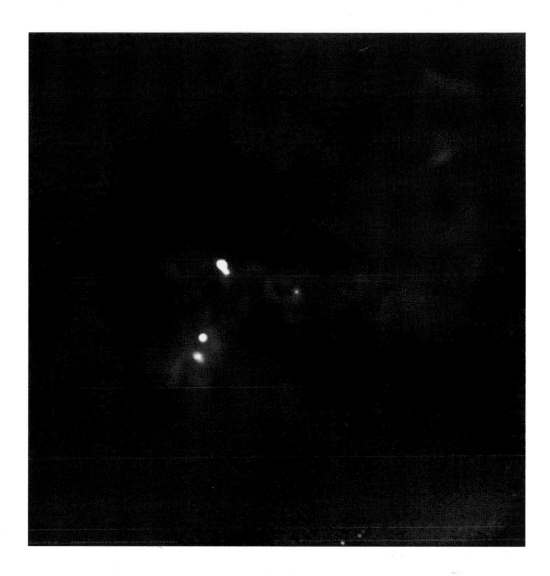

only form more massive stars, whereas cold clouds with high density also can form stars of low mass. The Jean's mass describes an idealized and very simple situation that ignores many other influences such as magnetic fields, and is at best an indicator of when collapse will occur.

The collapse of a cloud core is always non-uniform and starts in the densest part in the center. From there the collapse spreads outwards in an inside-out fashion. Basically the bottom falls out of increasingly higher layers in the core. Once set in motion, the gas is in free fall until it reaches the center, where a small stellar embryo of very high density begins to form. Therefore a protostar consists of a starlike central object containing a small fraction of its final mass, surrounded by a massive infalling envelope raining down.

Magnetic fields thread molecular clouds and create obstacles to star formation by resisting compression. Where magnetic fields resist compression, gas must diffuse past the field lines and fall to the self-gravitating center of a cloud core. This process (known as "ambipolar diffusion")[1] can delay cloud core collapse for millions of years. But in the absence of external influences, gravity will ultimately overwhelm magnetic cloud support. Once it starts, inside-out collapse proceeds only slightly more slowly than in the non-magnetic case.

Yet another roadblock to star formation comes in the form of rotation. Molecular clouds are in constant motion. As structure develops in a cloud and dense clumps and even denser cores form, the random gas motions can lead to the slow rotation commonly observed in the cores of molecular clouds. We can quantify the amount of rotation in such a core, or any other rotating object, by the *angular momentum* of the core, which is a measure of its mass, size, and spin.[2] In the absence of external torques, a rotating body will conserve its angular momentum.[3] The most common example of this is a spinning ice-skater. As she spins and pulls her arms towards her body, her rate of spin increases because her angular momentum is largely conserved. This is at the root of what is called the classical "angular momentum problem" of star formation: the angular momentum of a rotating cloud core is typically thousands of times greater than what can be contained in a star. If a typical core with a diameter of 0.3 light years and completing one rotation every million years were compressed to the size of our Sun, it would spin about its axis once every 10 seconds! The centrifugal forces would far exceed the gravitational force and the object would fly apart. Because of this imbalance, collapsing cloud cores must shed most of their angular momentum to form a star.

There are several ways a star forming cloud core can shed its spin. First, it can fragment into a binary or multiple star system whose orbital motion contains most of the core's angular momentum. Second, higher angular momentum material from the core will rain onto a spinning disk surrounding the tiny stellar embryo created by the infall of low-angular momentum gas. However to grow into a star, the embryo must accrete additional mass from the surrounding disk. This requires the further dissipation of orbital angular momentum within the disk. Magnetic fields trapped in the disk and torques exerted by self-gravitating clumps can transfer disk orbital angular momentum from the inner parts of the disk to its outer parts. As matter in the disk spirals towards its center, the outer radius of the disk must expand to conserve angular momentum. But the accretion of matter from the inner edge of the disk onto the equator of the protostar would still make the star spin too fast. The forming star's own magnetic field, perhaps generated by an internal dynamo, may be responsible for additional removal of angular momentum from matter accreting from the disk's inner edge onto the young star. As will be further discussed in Chapter 6, this stellar magnetic field may expel a fraction of the matter spiraling in from the disk's inner edge as powerful bipolar jets which carry away excess angular momentum, allowing low angular momentum material to settle onto the star. The interplay between angular momentum and accretion in a circumstellar disk is an essential driver for the evolution of the youngest stars, as we will discuss in Chapter 7.

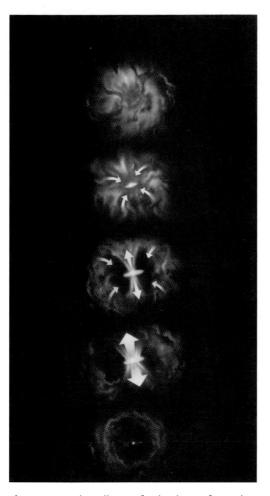

Figure 4.10. The collapse of a cloud core, formation of a disk surrounding a protostar, production of a bipolar outflow, clearing of the disk and formation of planets. (C.J. Lada).

The nature of stellar embryos has been the subject of detailed theoretical studies. The competition between gravity and pressure is won by gravity because the collapsing cloud core initially remains cold. This is because the infrared radiation emitted by the atoms, molecules, and dust grains in the collapsing core is able to escape through the infalling gas. The gas is said to be optically thin at this stage. As the cloud becomes denser, it starts to trap its own radiation and becomes opaque. The medium is then said to be optically thick. When the radiation can no longer escape, the central temperature begins to rise, raising the gas pressure. The optically thick core can then again resist gravity, and a new balance develops in the very innermost region of the collapsing core: a stellar embryo has been born. At this stage, the nascent star is called a hydrostatic core[4] because gravity and gas pressure has reached an equilibrium, as opposed to the highly dynamic situation in the infalling envelope.

The hydrostatic core is born with a small mass of about one-hundredth of a solar mass, but it is large compared to most stars, with a diameter of several Astronomical Units (AU), about the radius of the orbit of Mars. A powerful shock envelops the embryo as the infalling envelope continues to rain down on its surface. As the core rapidly grows in mass, its interior temperature rises to about 2000K, at which point the hydrogen molecules dissociate. As a result, the hydrostatic core becomes unstable again and collapses. This second collapse is extremely rapid, lasting only about 10 years, and is halted when most of the gas in the interior of the core becomes ionized. A second stable hydrostatic core then forms, this time with a size similar to that of the Sun. It too is surrounded by a shock where the rain of gas from the envelope impacts. This tiny but rapidly growing embryo is the true beginning of a new star, a protostar.

Observations of protostars

As a cloud core collapses to form a protostar and its surrounding disk, the infalling cloud remains cold with a temperature of about 10 degrees above absolute zero. But as material plunges onto the nascent star and its disk, the stellar embryo is heated and begins to glow at far-infrared and sub-millimeter wavelengths. The cocoon of infalling gas and dust becomes so dense that no optical or even infrared radiation can escape. As the first visual-wavelength light and near-infrared radiation produced by the forming star is absorbed by its surroundings, the temperature of the cocoon begins to rise. At this extremely early stage, only sub-millimeter observations can detect the presence of the budding star. An object in this stage is called a Class 0 source. Class 0 sources generally contain most of their mass in their infalling envelopes and disks, and only a small fraction in the stellar embryo. This is the first stage in the growth of a young star. Figure 4.10 illustrates the formation and evolution of a forming star and its disk.

Gradually, over tens of thousands of years, the temperature of the embryo, disk, and inner envelope rises. Eventually, the mean temperature of the dusty cocoon reaches 30 or 40 degrees above absolute zero. By then the warm dust emits enough far-infrared radiation to make the forming star detectable at far- and mid-infrared wavelengths. At this slightly more advanced stage the young star has accumulated more than half of its final mass, and is now called a Class I object.

Studies of regions of stellar genesis reveal that there are roughly ten times as many Class I sources as Class 0 objects. If one evolves into the other, the Class 0 stage must be ten times shorter than the Class I phase. However, estimating the duration of these early evolutionary stages is difficult. Although brief on astronomical time-scales, they last much, much longer than a human lifetime. The best current estimate of the duration of the Class 0 phase is about ten to twenty thousand years. The longer-lasting Class I stage is probably a few hundred thousand years in duration.

Disks surrounding highly embedded and obscured protostars can only be seen at radio and sub-millimeter wavelengths. But a single-dish radio telescope pointed towards a protostar also detects signals from the surrounding cloud core and molecular cloud, making it difficult to distinguish the emission from a disk. Even in the nearest protostars located in the constellation Taurus, disks have very small angular sizes. The largest disks may be 1000 AU across and subtend angles of only 7 arcseconds in Taurus or 2 arcseconds in Orion: much too small to resolve with an ordinary radio telescope. The study of disks requires radio telescopes with resolutions of about an arcsecond. By operating a number of radio telescopes in concert as an interferometer, astronomers can obtain the resolution necessary to directly image circumstellar disks. These radio maps can be used to measure the sizes, masses, compositions, grain properties, and orbital motions of circumstellar disks.

Some Class I sources have depleted their surrounding envelopes sufficiently to allow near-infrared and sometimes even optical radiation to escape. Protostars in this stage of development can often be observed at these wavelengths unless their disks are viewed edge-on, in which case a bipolar reflection nebula is seen. The cameras on the Hubble Space Telescope have produced many images of disks surrounding young stars in this evolutionary stage (Figure 4.11). The light has been scattered by material in the circumstellar environment, and the circumstellar disks are seen in silhouette against this illuminated background. When the disk does not obscure the central star, its blinding light can render the disk invisible.

Though envelopes and disks are readily detected in some cases, measurement of their infall motion and rotation is very difficult. These motions are small, amounting to only about a kilometer per second at a distance of a thousand AU from a solar mass young star. Interferometric observations have detected both infall motions in the envelopes of newborn stars and the rotation of their disks. These observations show that gas along the line of sight

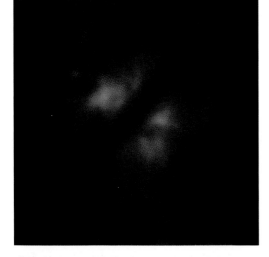

Figure 4.11. An infrared image obtained by the HST showing a dense circumstellar disk surrounding the young source IRAS 04302+2247 in the Taurus clouds. (D. Padgett/STScI/NASA).

towards a protostar often exhibits small motions away from us and towards the star; gas on the far side moves towards us and towards the star. This is the expected behavior of an infalling envelope. Also, the Doppler shifts of emission produced by molecules have provided direct measurements of the orbital motions of disks.

The initial mass function

Stars come with a variety of masses. As will be discussed in Chapter 9, stars with masses larger than about 100 times that of the Sun would be unstable because their furious output of radiation would lead to the ejection of their outer layers. The lower limit is around eight percent of the mass of the Sun, the minimum mass needed to reach the pressure and temperature required for the thermonuclear burning of hydrogen. Less massive objects are called brown dwarfs.

Observations show that massive stars are relatively rare, and that stars become progressively more common for smaller masses; the Universe is teeming with low-mass stars shining with feeble light. The distribution of stellar masses is known as the stellar *Mass Function*. It is, however, not really representative of the numbers of stars of different masses born in a cloud, because stars lose mass during their lives and especially because stars of different masses have different life spans. As mentioned earlier, the most massive stars live for only a few million years, while stars with masses less than 0.8 times the Sun's mass that formed right after the Universe was born are still around. If we correct for these effects, we can derive the *Initial Mass Function*, which is the distribution of masses produced by molecular clouds.

Why are some stars born with low masses and others with large masses? Is the final mass determined by the mass of the core from which it forms? Can a young star accrete all of the mass in such a core? Observations and theoretical models show that stars forming in more massive and denser cores tend to grow faster and gain more mass. Indeed, the mass distribution of pre-collapse cloud cores is similar to the Initial Mass Function of stars, implying that the stellar mass distribution is determined by the fragmentation processes that produce cloud cores.

However, this cannot be the whole story. As discussed in the next chapters, random motions of stars and gas, outflows, interactions with sibling stars, heating and evaporation of the parent cloud by newborn stars, magnetic fields, and the influence of external radiation, winds, and explosions may limit the amount of mass any given protostar can accumulate from its environment. Furthermore, the conservation of angular momentum implies that as some mass falls onto the star, other matter must be expelled. So, if the parent core has any spin, all of its mass cannot end up in the central star. It appears likely that a typical stellar mass is several times smaller than the mass of the cloud core out of which it formed.

Energy sources of protostars

All stars emit light. While there are several possible sources of energy that could produce the light emitted by a star, only one energy source can sustain stellar energy output over its lifetime, namely thermonuclear fusion. Stars are natural thermonuclear reactors which fuse hydrogen and other light elements into heavier elements. Hence, a star is *defined* as a celestial body that derives its light from nuclear burning.

Protostars get most of their energy by contracting. After the formation of the first and second hydrostatic cores, the stellar embryo settles into a quasi-stable state. But newborn stars need energy to keep hydrogen dissociated and ionized, to maintain their gas pressure high enough to avoid further collapse, and to radiate at their surfaces. All this they can achieve by an imperceptible contraction that continuously releases gravitational energy. Obviously a young star cannot contract forever, but the length of time that a star can maintain its energy needs by contraction (known as the Kelvin-Helmholtz timescale) is more than sufficient for its needs during stellar youth.

Hydrogen has two stable, long-lived isotopes; ordinary hydrogen whose nucleus consists of a single proton, and deuterium whose nucleus contains a proton and a neutron. Deuterium burns to helium at about one million degrees, while hydrogen burns at about ten million degrees. So as the cores of the young stars heat up past one million degrees, they burn deuterium. Since there is only one deuterium atom for every hundred thousand hydrogen atoms, this is a minor process compared to the later burning of ordinary hydrogen. But deuterium burning produces an amount of energy comparable to that gained by contraction.

When do stars begin to burn their nuclear fuel? When does a star become a star? By definition, this happens when a star reaches the *main sequence*, as discussed in the following.

The two most easily derived characteristics of a star are its light output, or luminosity, and its surface temperature. We can plot these stellar parameters against each other in what is called a *Hertzsprung-Russell* (HR) diagram,[5] named after the Danish astronomer Ejnar Hertzsprung and the American astronomer Henry Russell, who first investigated the relation between these two observables. The large majority of normal stars fall in a band that runs diagonally through the diagram, with the hottest and most luminous stars at one corner, and the coolest and least luminous at the other. This band is known as the stellar main sequence.

Stars are not born on the main sequence. A protostar first appears in the H-R diagram as a low-luminosity and cold object. As the stellar embryo evolves, it warms and becomes more luminous. Accreting protostars derive most of their luminosity from the release of gravitational potential energy by infalling matter. The kinetic energy of infall is converted into radiant energy at the accretion shock where this matter impacts the hydrostatic core. For objects destined to be low-mass stars, the luminosity of matter accreting onto the stellar embryo during the Class 0 and Class I phases can be orders of magnitude higher than the main-sequence luminosity of the final star. Thus, low-mass protostars are cool, overluminous objects.

A young star reaches the main sequence when it starts burning ordinary hydrogen into helium at a steady rate. To reach this stage may take anywhere from a few million years for several-solar-mass stars to about 100 million years for the lowest-mass stars near the brown dwarf limit. This is still very short compared to the many billions of years that low-mass stars can exist as stable main-sequence stars.

Between the protostellar and main sequence stages, young stars are known as *pre-main sequence* stars. The protostellar stage ends when the stellar embryo has gained more than half of its final mass. This transition nearly coincides with the emergence of the first light from the newborn star. When placed in the Hertzsprung-Russell diagram, these freshly emerged pre-main sequence stars define a locus called the *birth line*.[6]

The energy produced in the stellar interior diffuses to the stellar surface. It can do so in two ways: by large-scale convective motions of the gas in huge eddies, or by radiation. A newborn star starts out fully convective from its surface to its very center. Convection brings fresh deuterium to the center, where it burns to helium as soon as contraction raises the central temperature to a million degrees. Deuterium burning has the unusual property that it acts as a thermostat and does not allow the central temperature to rise above one million degrees. So even though the contraction of the star would produce higher central temperatures, this is not possible until the deuterium burning stops. During this stage, called the Hayashi track[7] in the Hertzsprung-Russell diagram, the star then becomes smaller and smaller and less and less luminous. But at a certain point the energy transport in the stellar interior begins to switch from convective to radiative. As a result, there is no longer a steady supply to the interior of fresh deuterium from the higher layers, and the deuterium burning ceases. Once the deuterium thermostat is switched off, the central temperature begins to rise as the star continues to contract. The surface temperature of the star also begins to rise, and even though the star becomes smaller, its total luminosity stays approximately constant. Eventually the star becomes sufficiently hot in its interior to start nuclear burning of ordinary hydrogen as the star reaches the main sequence. Stellar youth is over.

5 Companions in birth: binary stars

Binary stars

As astronomers of the eighteenth century examined the sky through their ever-improving telescopes, they noted that stars frequently appear double. Initially, pairing was interpreted as chance alignments along the line of sight. But as the number of doubles increased, it became clear that at least some doubles are physical pairs. Through painstaking observations over decades, William Herschel realized that some pairs orbit each other, proving their physical association. In 1802, Herschel coined the term *binary stars* to describe "the union of two stars, that are formed together in one system, by the laws of attraction."

Because our Sun is single, for many years it was taken for granted that most other stars are also single. However, we now recognize that multiplicity is the norm among stars. Roughly two-thirds of the stars in the sky are binaries.

Figure 5.1 shows an image of the famous binary star Albireo in the constellation Cygnus. The two stars can be easily separated in small telescopes. Colors of stars are normally not so easy to perceive, but because the two components of Albireo have very different surface temperatures and they are seen right next to each other, their color difference is striking. The orange hue of the brighter star shows that it is cooler, and the blue color of the fainter component reveals it as a hot star.

Binaries have a wide range of properties. The complex interplay between stellar evolution and orbital dynamics can result in a plethora of fascinating and bizarre phenomena. The stars in some binary systems complete an orbit in less than an hour. In such systems, the stars are so close that they may physically touch each other. Very close systems can exchange mass. As the components of a close binary age, their reservoirs of hydrogen dwindle. As a result, the more massive member of such an aging, close binary swells into a red giant star,[1] which can pour much of its mass onto the companion. Eventually, the secondary star may itself swell, and return some of its mass onto the remnant of the primary, which by now may be a white dwarf, a neutron star, or even a black hole.[2] If a companion star explodes, the surviving star may be released from the gravitational bonds of the binary and fly away with its full orbital speed of hundreds of kilometers per second. Mass exchange in short-period binaries is responsible for phenomena such

Figure 5.1. The binary Albireo in the constellation Cygnus consists of two stars with very different surface temperatures, resulting in a striking contrast in color. (Johannes Schedler).

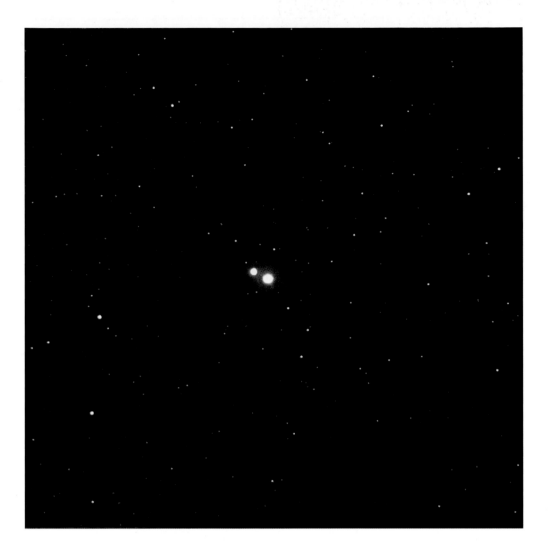

as cataclysmic variability, pulsating X-ray sources, violent stellar explosions, and powerful outflowing jets.[3] At the other extreme, the stars can be so far apart that they take millions of years to complete a single orbit around each other. The most common separation between members of binary systems is about 30 AU and it takes the stars in such systems anywhere from decades to centuries to complete an orbit.

The detection of binary and multiple star systems requires observational techniques tailored to their separations. Widely separated systems can be directly resolved into their individual components if they are not too far away from us. But in most binaries, the stars are too close to each other to be resolved into an image. Indirect methods must then be used to observe such systems. We can discover their binarity by analyzing their blended light. The orbital speeds of close binaries have velocities about their common center of mass that range from a few to well over a hundred kilometers per second for the tightest systems. Each star's spectrum will contain distinct features produced by the absorption of various elements or ions. The stellar motions will shift their wavelengths back and forth with the orbital period, revealing binarity through the Doppler effect. Such *spectroscopic binaries* are quite common. Over 15 percent of the stars in the sky reveal their companions in this manner.

If our line of sight lies in the orbital plane of the two stars in a binary, the stars will regularly pass in front of each other, periodically diminishing their light. Such *eclipsing binaries* are rare because it is unlikely for their orbital planes to be close to our line of sight. Only about one in a thousand stars show such eclipses. Nevertheless, eclipsing binaries are important because their light curves can be combined with their Doppler velocities and orbital periods to determine the radii, masses, and temperatures of the individual components. The best studied eclipsing spectroscopic binary systems have yielded these fundamental stellar parameters to an accuracy of better than one percent. The detailed analysis of binary systems have provided critical data about the basic properties of stars.

Although in some binary systems the stellar masses are similar, in most cases they are quite different. In such systems the more massive star is generally much brighter than its companion. In those cases, it can sometimes be difficult to detect the dimmer companion in the glare of the brighter star, either visually or spectroscopically. Many faint companions have been missed, and the actual fraction of binary stars may in fact be larger than two-thirds. Recently, observers have started to combine observations at visual wavelengths of the brighter and hotter component with infrared observations of the fainter and cooler companion, allowing each to be observed at the wavelengths where they are best detected.

Some systems contain more than two stars. About ten percent of binaries have such additional companions. An example is the brightest star in the southern constellation of Centaurus, known as Alpha Centauri, which for many years was believed to be the closest star to the Sun. It takes the light from Alpha Centauri only four years to traverse the distance to us. More detailed observations show that it is not a single star, but a system of three. Two stars are similar to our Sun and orbit each other with a separation of about 20 AU, roughly the distance between the Sun and Uranus. Additionally, a faint red dwarf star called Proxima Centauri is located about a hundred times farther out, slowly orbiting the pair of bright stars. Proxima is a little closer to us than Alpha and is therefore the nearest star to the Sun.

The configuration of the Alpha Centauri triple system, with two close components and another distant one, is typical for triple systems. Such configurations are called hierarchical, and calculations show that only hierarchical systems can survive for a long time. Higher-order multiple systems must also be hierarchical. Quadruple systems, for example, usually contain two pairs of close binaries that are well separated from each other.

Young binaries and multiples

Over the last decade, searches for binaries among young stars have been conducted at both optical and infrared wavelengths. Numerous young binaries with ages less than 10 million years have been found. Surprisingly, there are more binaries and multiples among young stars than among older stars. In relatively isolated star-forming regions such as Taurus, studies show that as many as 90 percent of the young stars are binaries or multiples. On the other hand, in the rich young clusters such as Orion's Trapezium cluster, the percentage of multiples is similar to that found in the solar vicinity, namely about two-thirds. Either most stars in the sky were born in Orion-like environments, or the mortality of young multiples is high.

One possible explanation for the excess numbers of multiples in some star-forming regions is that they have wider separations while young, and are therefore easier to detect. But when we compare young systems with older binaries in the same separation ranges, we again find more young binaries, suggesting that the higher binarity among young stars could be real. Variations in binarity among different star forming regions have also been noted and could be explained by interactions between the stars. In crowded regions like Orion, young stars have a greater probability of passing close to each other. Mutual interactions can lead to the destruction of some binaries, or a faster shrinkage of their orbital separation. But it is also possible that some star forming regions produce more binaries than others, and that the fraction of binaries among field stars represents an average over different formation environments.

How do circumstellar disks behave in young binary systems? Disks typically have radii of the order of 100 AU. In binaries with much larger separations, disks around the individual stars may remain unperturbed by the distant companion. Similarly, if the binary components are very close, the system may be surrounded by a *circumbinary* disk which would behave as if orbiting a single star. But when the binary components have separations of the order of 10 to 100 AU, there is little room for disks. Disks around the individual stars are then constrained to be only a few AU in diameter, and circumbinary disks would have inner radii of order a 1000 AU. Intermediate sized disks would suffer strong tidal deformations, warping, and collisions that would lead to their rapid destruction. Observations show that binaries of all separations have disks. But observations also confirm that disks associated with binaries having separations between 10 and 100 AU contain much less material than those around isolated stars; such systems are unlikely to have or to form planets.

Numerical modelling can explore the important effects that companions have on a circumstellar disk. Figure 5.2 shows four panels in a simulation of a star surrounded by a circumstellar disk which is approached by and then penetrated by a diskless companion. As the companion ploughs through the disk, it sets up major waves and temporary spiral features. Repeated passages like that eventually truncate the disk.

Visual or near-infrared observations of binaries among the deeply embedded Class 0 and Class I sources are difficult due to high extinction. Although longer-wavelength infrared observations can penetrate their dusty

Figure 5.2. A simulation of a stellar companion penetrating the circumstellar disk of a young star. (N. Moeckel).

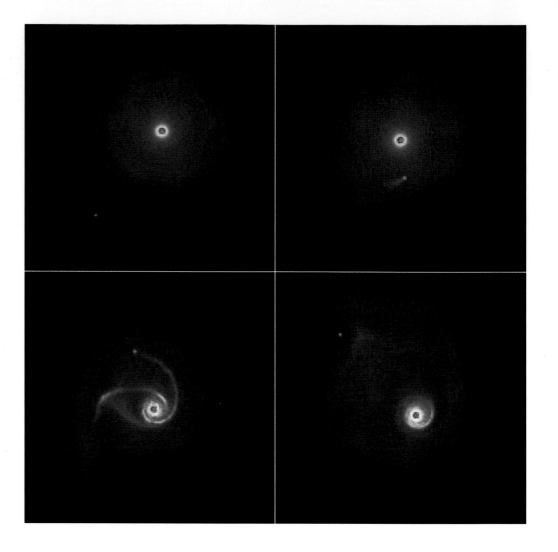

environments, the best way to observe embedded multiple stars is to use centimeter wavelength facilities such as the *Very Large Array* in New Mexico. Very young stars often emit at radio wavelengths, permitting researchers to probe embedded stars with the high spatial resolution offered by radio interferometers. Figure 5.3 shows an infrared image obtained with the Hubble Space Telescope of the young embedded source L1551-NE. The image shows a large outflow cavity illuminated by a source that is not detected even at near-infrared wavelengths. But Figure 5.4 shows a VLA radio interferometer image in which the source is clearly detected. It is a close binary with a separation of about 40 AU in the plane of the sky.

The birth of binaries

Since about two-thirds of all stars are in binary systems, our understanding of star formation would be seriously incomplete if we could not explain how binary and multiple stars are born. Over the years a number of theories of binary formation have been put forward.

Figure 5.3. An HST near-infrared image of the embedded source L1551 NE. A cavity opens up from the newborn star. (B. Reipurth & J. Bally, NASA/STScI/HST).

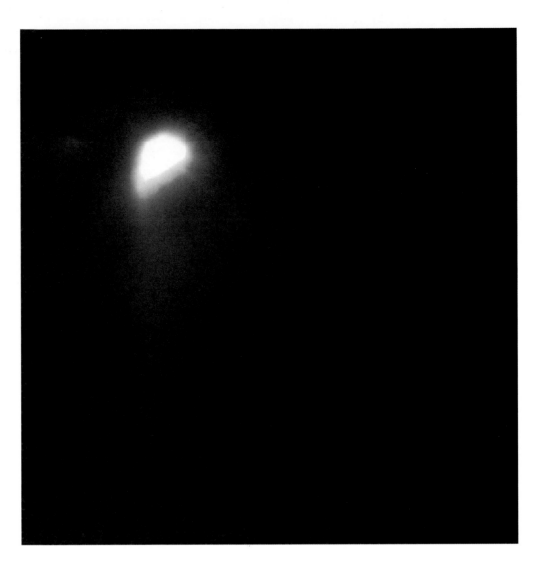

L1551NE

B

A

Figure 5.4. A radio image of L1551-NE obtained with the Very Large Array radio telescope shows that the source is a binary. Tickmarks are 0.5 arcsec apart. (B. Reipurth & L.F. Rodriguez, NRAO/VLA).

The earliest idea, dating from the nineteenth century, suggests that as a star moves around in the Milky Way, it can occasionally pass very close to another star and *capture* it to form a binary. However, we now know that average distances between stars, once they have left their region of birth, are so large that such encounters are exceedingly rare. Even if two stars were to meet, it is not possible for them to go into orbit around each other without shedding some kinetic energy.[4] To become bound, a third body that can absorb this energy has to be present. Therefore, this idea has been abandoned.

Another possibility is that a rapidly spinning protostar deforms into a bar-shaped configuration as it contracts and eventually splits into two bodies. Such *fission* of a protostar would only produce very close binaries and could not explain the many very wide systems. Furthermore, computers have performed detailed calculations of protostellar behavior as a star contracts, and fission does not occur in any realistic models.

A viable mechanism was proposed by the British astronomer Fred Hoyle in the 1950s. He demonstrated that as a cloud contracts in the early phases of

Figure 5.5. A computer simulation of the collapse of a prolate cloud shows how fragmentation leads to the formation of a binary star. (Alan Boss).

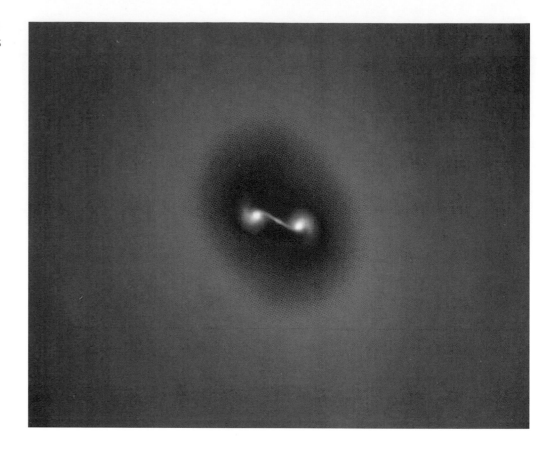

star formation, it may break up into several clumps. Calculations show that such *fragmentation* is more likely in cloud cores that are initially elongated. There is now general agreement that fragmentation is an important process in cloud collapse and that it is the most likely mechanism to form wide binaries (Figure 5.5).

While fragmentation can produce wide binaries, it cannot produce close binaries easily. As a core collapses, its central density and temperature increases. However, hot gas is relatively more stable to small scale fragmentation than cold gas. This feature tends to inhibit fragmentation into binaries with separations of about 10 AU or less. But at least 20 percent of binaries have companions closer than 10 AU, so another mechanism is required to explain their existence. As we shall see, such close binaries can be made by shrinking the orbits of pairs of stars in small multiple systems.

Disintegration of multiple stellar systems

Surveys of star forming regions show that it is rare for stars to form in isolation. The large majority of stars form in groups ranging from a few members to thousands of stars, with ensembles of about 100 stars being common. In such small clusters, stars are likely to interact gravitationally with other members. Thus, dynamical processes and interactions may play important roles in the early lives of stars.

In a system of two bodies, orbits can be predicted for any initial configuration with great accuracy and far into the future. But there is no general solution for the evolution of non-hierarchical configurations of three or more bodies, their motions can be chaotic and unpredictable over long periods. But this complex behavior of three bodies can be explored in great detail with computers. The behavior of three bodies responding to their mutual gravity can be divided into three categories. During *interplay*, the three members execute random motions with no periodicity. Two members, often the two most massive, may form a temporary binary which is continuously perturbed by approaches of the third member. During a rare *close triple approach*, the three bodies are simultaneously brought close together. During such three way encounters, energy can be exchanged between the components. The third class of motion is called *ejection*, since one of the three members, usually the least massive one, is kicked out of the triple system. Ejections can occur only following a close triple approach during which a tight binary is formed that releases gravitational potential energy that is absorbed by the third member, thus loosening its ties to the system. Ejection transforms the triple into a close binary and an ejected member which will either eventually fall back towards the binary, or lead to complete escape. If the third member remains loosely bound to the binary, it forms a stable hierarchical triple system, as is commonly observed among stars. Such interactions therefore mark the transition from chaotic non-hierarchical to stable hierarchical systems. Computer models of the evolution of non-hierarchical systems show that they tend to reconfigure into hierarchical multiples within about a 100 revolutions of the stars around each other.

When the third member is ejected, the remaining binary must become more tightly bound to conserve energy. Even though the three members may initially have been relatively far apart, the final outcome is a pair of stars in a close binary and either a loosely bound or an escaping member.

Similar interactions take place in systems with 4, 5, or more stars, in which member after member is ejected, leaving behind one or more tight binaries. Thus, the disintegration of small multiple systems is widely seen as the mechanism that can form close binaries.

Competition in small stellar systems

Numerical simulations with powerful computer clusters allow theoreticians to follow the detailed collapse and fragmentation of an unstable cloud core. Such hydrodynamical calculations are highly complex if carried out in three dimensions and in time, involving gravity and sometimes magnetic fields. One technical hurdle is related to the fact that the density of a collapsing cloud core will shrink by more than 20 orders of magnitude, so a resolution that is sufficient at the beginning of the calculation is woefully inadequate towards the end. New codes overcome some of these technical problems and

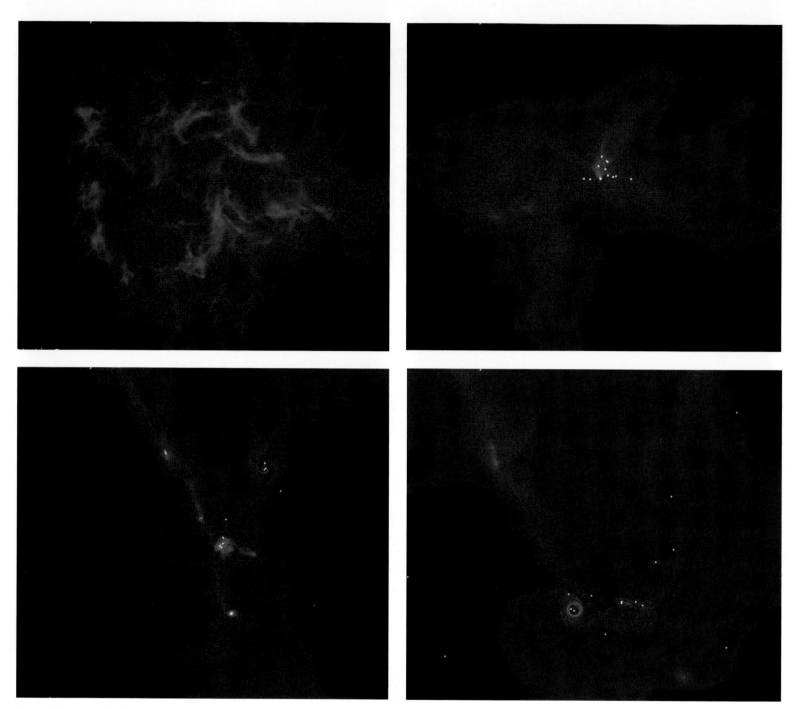

Figure 5.6. These four images show a computer model of the birth of a small group of stars from the collapse of a dense turbulent cloud core. (Matthew Bate/UK Astrophysical Fluids Facility).

simulations now give fairly realistic pictures of the fragmentation of a cloud core into stars and the evolution of the resulting clusters.

These simulations show that accreting stars compete with each other for mass from the infalling envelope. Accretion rates are unequal, because stars accrete more when they are close to the dense center of the infalling envelope. In a "the rich get richer, the poor get poorer" scheme, the more massive a star, the more it accretes. Furthermore, more massive stars tend to sink to the cloud center where they encounter higher accretion rates because of the higher density. Stars that started with less mass, or ones far from the center will remain less massive since they accrete less. In the complex dynamical interplay between sibling stars in multiple systems, the smaller members will have a higher probability to be flung around by their heavier brethren. Eventually, some of these smaller stars will be completely ejected from the cloud core.

Figure 5.6 shows a sequence of images taken from a three-dimensional computer simulation of star birth in a 50 solar mass cloud core. The turbulent core collapses into sheets and filaments. As stars form in the densest parts about two hundred thousand years after the beginning of the simulation, they mingle and undergo a cosmic dance. Several small multiple systems of stars are formed, and some of the lowest mass members are ejected from the nest. Pairs of the stars remaining in the systems enter into eccentric orbits around each other, forming binary stars.

In case of fierce competition, the smallest stellar embryos will be kicked out of the cloud so soon after birth that they have too little mass ever to ignite the thermonuclear burning of hydrogen which sustains stars. Such runts of the litter will become brown dwarfs floating among the stars of the Milky Way.

6 Outflows from young stars

The objects of Herbig and Haro

In the late 1940s, a young student, George Herbig, was working on his doctoral dissertation at the Lick Observatory in California. He was studying regions in Orion which harbored many faint, variable stars, already suspected of being much younger than the rest of the stars in the sky. Simultaneously, Guillermo Haro, a Mexican lawyer-turned-astronomer, was also investigating Orion from the Tonantzintla Observatory in Mexico in search of more young stars. Both astronomers noted some strange small blobs with peculiar spectra on their photographic plates. Herbig and Haro knew that stars produce continuous spectra, with dark lines superimposed due to absorption by atoms and ions.[1] However, unlike anything seen so far in star-forming regions, these new objects showed only a set of isolated intense lines of emission in their spectra. Herbig's and Haro's objects were not quite stellar in appearance; rather they were slightly fuzzy, as if they were small compact nebulae. Their location in regions near dark clouds and young stars suggested that perhaps these objects might be in a transitional phase between a cloud and a star. Perhaps these nebulae traced the birth of young stars.

For many years these objects, now known as Herbig-Haro objects, remained an astronomical enigma. But about a quarter of a century ago, two discoveries provided the keys to solve their mystery. First, Richard Schwartz of the University of Missouri realized that the spectra of Herbig-Haro objects perfectly fit what would be expected from an interstellar shock wave. Shocks are produced when gases collide with speeds larger than the speed of sound. Shocks heat the gas and cause it to emit copious amounts of radiation similar to what is seen in Herbig-Haro objects. Second, George Herbig and co-workers discovered that these peculiar nebulae are not fixed in the sky, but move with speeds of several hundreds of kilometers per second. And when tracing their paths back in time, they found that Herbig-Haro objects originate from very young stars still deeply embedded in their parental clouds. Herbig-Haro objects were sometimes found in pairs, located symmetrically on each side of a young star and moving away supersonically. Figure 6.1 shows a region about two degrees south of the Orion Nebula that contains many young stars and Herbig-Haro objects. Figures 6.2 and 6.3 show the first outflow to be recognized, the bright Herbig-Haro objects called HH 1 and 2.

Figure 6.1. A large cloud region with numerous Herbig-Haro flows and active star formation located south of the Orion Nebula. (T.A.Rector/NOAO/AURA/NSF).

Figure 6.2. The HH 1 flow as seen by the Hubble Space Telescope in the light of hydrogen and ionized sulfur. (B. Reipurth & J. Bally, NASA/STScI).

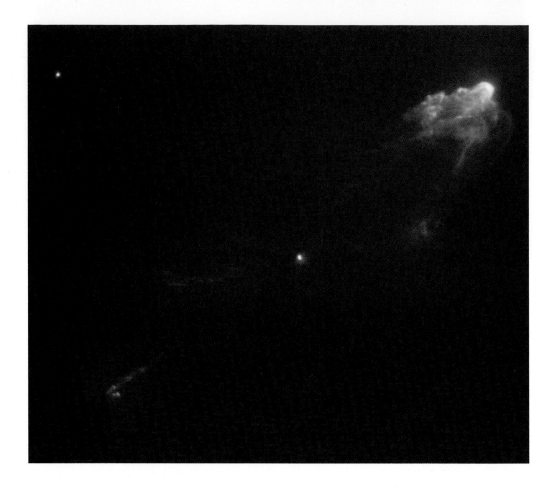

Figure 6.3. The HH 2 flow as seen by the Hubble Space Telescope in the light of hydrogen and ionized sulfur. (B. Reipurth & J. Bally, NASA/STScI).

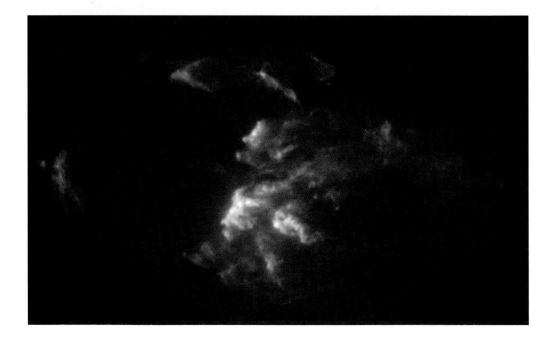

The nebula in the upper right of Figure 6.2 has the appearance of a bow or the nose of a blunt bullet. The lower left portion of Figure 6.2 shows a jet of high velocity gas streaming towards HH 1. The source of the jet is a hidden multiple star system. Figure 6.3 shows HH 2 which has broken up into a number of small bows. Both objects move with speeds of several hundred kilometers per second away from the young star system located midway between them.

After half a century investigating Herbig-Haro objects, we now know that they are shocks in extremely fast flows which are ejected from very young stars in a bipolar fashion.[2] Their discovery was a major surprise to astronomers, who expected that young stars would show only infall motions as they gradually built up their masses. Nobody predicted the discovery of outflow motions. Detailed studies of Herbig-Haro objects have provided unique insights into the strange behavior of the very youngest stars.

Jets from newborn stars

In the southern sky along the Milky Way lies the constellation of Vela. This region contains many Bok globules: small, very compact and dark clouds that float like islands in the space between the stars. Because it is associated with Herbig-Haro objects, one of these globules has attracted much attention (Figure 6.4). The Bok globule has a windswept appearance, indicating a flow from the upper right of the picture. Several very massive stars, which are flooding their surroundings with intense light, are located in that direction. This radiation produces the cometary shape of the globule and is also responsible for its luminous skin. Additionally, irradiation may have compressed the globule, causing it to collapse, and thus triggering the globule to give birth to a young star.

A young star embedded near the upper edge of the globule was discovered when the IRAS satellite made its far-infrared survey of the entire sky. Because it is still wrapped in its placental material inside the cloud, it cannot be seen at visible wavelengths. But radiation from the embedded star heats its surroundings and makes the region easily detectable at infrared wavelengths.

Although the star is not visible, its existence is revealed by the Herbig-Haro objects it is expelling. A bright, collimated stream of gas flows towards the upper left of Figure 6.4. This Herbig-Haro flow, known as HH 47, is actually bipolar, but the opposite lobe, flowing away towards the lower right of the figure, is burrowing deeper into the globule and is therefore mostly invisible. Only the very tip of this counterflow, placed symmetrically around the young star with the tip of the main flow, can be seen protruding behind the edge of the globule.

At infrared wavelengths, the picture changes dramatically. Figure 6.5 shows an image obtained by the infrared *Spitzer Space Telescope*. The dark cloud is now mostly transparent, showing the background stars that were obscured

Figure 6.4. The HH 47 jet emanates from a newborn star embedded in a dense Bok globule in the southern Milky Way. (B. Reipurth).

in the optical image. The newborn star that drives the jet is the blindingly bright source near the edge of the cloud. Furthermore, the infrared image clearly shows the embedded counterflow to the HH 47 jet as a long tube lined with luminous gas where the outflow from the young star is penetrating the globule.

The Hubble Space Telescope also observed HH 47, and the resulting image is shown in Figure 6.6. This is one of the best images ever obtained of a Herbig-Haro jet. It reveals the intricate details of the flow as it plows through the ambient medium. The jet produces luminous shocks as it hits its surroundings and as faster material along the flow axis overtakes and collides with slower gas. By taking spectra of the jet, astronomers can determine its velocity along the line of sight. Additionally, images taken several years apart clearly reveal the movement of the gas parcels. From these data, astronomers

Figure 6.5. The HH 47 jet as seen by the Spitzer Space Telescope in infrared light. The source is here very bright, and a luminous cavity is evident around the counterflow which is invisible at optical wavelengths. Note that the optical and infrared images are oriented slightly differently. (NASA/Spitzer).

have found that the jet is flowing towards us at an approximate angle of 30 degrees to the plane of the sky with a speed of no less than 200 km/s. The extent of the jet on the sky together with its distance of about 1500 light years tells us its size, which is about half a light year. The flow has taken several thousand years to reach its present size.

Even better collimated HH jets have been found in Orion's star forming clouds. Figure 6.7 shows HH 34, south of the Orion Nebula. A newborn star is becoming visible as it digs its way out of its small parent cloud. It emits a powerful jet that contains numerous small, luminous knots where fast flow components catch up and shock against slower ejecta. Further along the flow axis, towards the lower left of the picture, there is a spectacular bow shock. This represents the end point of an earlier episode of activity of the driving source. Symmetrically placed about the source there is a counter bow shock. No counter jet is seen, perhaps because it is shrouded from view by a cloud of gas and dust.

HH 111 is another highly collimated jet shown in Figure 6.8 in an image obtained with the Hubble Space Telescope. The right part of the figure shows an optical image of HH 111 emerging from a dense cloud core. No trace of

Figure 6.6. The HH 47 flow as seen by the Hubble Space Telescope in the light of hydrogen and ionized sulfur. (B. Reipurth & J. Morse, NASA/STScI).

Figure 6.7. The HH 34 jet is highly collimated and shows a bipolar structure. (B. Reipurth & J. Bally).

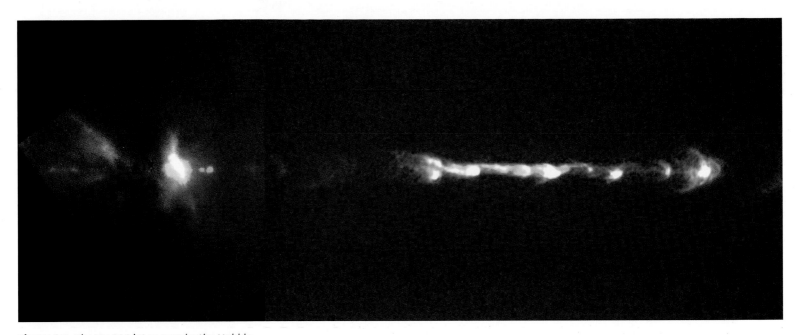

Figure 6.8. The HH 111 jet as seen by the Hubble Space Telescope at visible wavelengths (right part) combined with an image of the source region at infrared wavelengths (left part). (B. Reipurth & S. Heathcote, NASA/STScI).

the embedded energy source can be seen at optical wavelengths. The left part of the figure shows an infrared image that reveals the chaotic region around the driving source, visible as a deep-red star precisely on the jet axis. Luminous patches trace a cavity dug by the outflow. The jet lies on the axis of a large, dusty disk that surrounds the source and which is seen in silhouette against these patches. The total extent of the jet is about one-third of a light year, typical for collimated Herbig-Haro jets.

Herbig-Haro jets constitute a finely collimated subset of the Herbig-Haro objects. There are speculations that only the very youngest stars have the ability to produce long, highly collimated jets. Somewhat more evolved objects may gradually lose this ability, producing instead more and more broken-up and diffuse shocks. But there are puzzling exceptions to this, which indicates that our understanding of the formation of Herbig-Haro jets is still limited.

Herbig-Haro jets often show chains of more or less evenly spaced bow shocks symmetrically placed around their sources. This suggests that the young stars repeatedly undergo cycles of outflow activity, an important insight that tells us much about the way very young stars build up their mass, a point to be discussed in detail later.

How large can Herbig-Haro flows become? It has recently been recognized that they can attain gigantic proportions, many light years in length. The HH 111 jet drives one of the largest flows known. Wide-field images show that, in addition to the jet seen in Figure 6.8, a large number of bow shocks are located farther out on either side of the source, with each lobe stretching more than 10 light years from the star! To grasp the enormity of this distance, recall that the nearest star to the Sun is four light years away. That is, the powerful jets produced by newborn stars can extend several times the typical separation between stars in our Milky Way.

Churning dark clouds

Herbig-Haro objects are one of several outflow phenomena found in star forming clouds. Soon after the technique of observing molecules at millimeter wavelengths was perfected in the 1970s, studies of cloud structure and motions indicated that certain localized regions in star-forming dark clouds have velocities much larger than the mean velocity of the cloud. These regions were found to be bipolar, with one lobe approaching and the other lobe receding, and always centered on a young star. These objects are now known as bipolar molecular outflows.

Molecular outflows are massive, often containing many solar masses of moving material. So much matter could not have been spewed out by a young star which in most cases has a mass less than the Sun. Molecular outflows usually have much lower velocities than Herbig-Haro flows, typically only 10 to 20 km/s, compared to 100 to 500 km/s for the HH jets.

What is the relation between HH jets and molecular outflows? Molecular outflows consist of material from the ambient molecular cloud which was entrained and accelerated by Herbig-Haro jets as they plow through their parental clouds. The fast, but not so massive Herbig-Haro flows consist of material ejected from the immediate surroundings of a young star. This material rams into ambient gas, transferring momentum to the cloud. As the Herbig-Haro flows slow down, the stationary gas is accelerated, creating the bipolar lobes of high velocity molecular gas around young stars.

Figure 6.9 shows a millimeter map of carbon monoxide (CO) emission from the L1551 cloud, which contains only about 40 solar masses of gas. Greenish areas show quiescent gas from the cloud, and the blue and red regions show the lobes of high velocity molecular gas associated with several molecular outflows. Young stars in the cloud are marked as red dots.

Figure 6.10 shows an optical image of the L1551 cloud covering approximately the same area of the sky as the millimeter image of Figure 6.9. Many Herbig-Haro shocks are visible, and comparison of the two figures shows the impact of fast Herbig-Haro flows on the surrounding CO emitting gas. This low-mass cloud has spawned about a dozen young stars. Nearly half drive overlapping outflows into their surroundings. The two most luminous of these young stars power the brightest HH objects and the most massive bipolar molecular outflows. Some shocks in this region lie along lines of sight so clear that background galaxies can be clearly seen. In some cases shocks are found well beyond the spatial extent of the main bipolar CO outflow seen at millimeter wavelengths, because the Herbig-Haro flows have punched completely out of the cloud to regions where there is no longer any CO gas that can be entrained and set in motion.

In addition to the two brightest infrared sources, known as L1551 IRS5 and L1551 NE, which drive the main molecular outflows, there are several other

Figure 6.9. A CO map of the L1551 cloud shows a bipolar molecular outflow. Quiescent gas is pale green, high-velocity blueshifted gas is blue, and redshifted gas is red. A newborn star is embedded between the red and blue lobes. Red dots mark the location of young stars. (J. Bally & Y. Billawala).

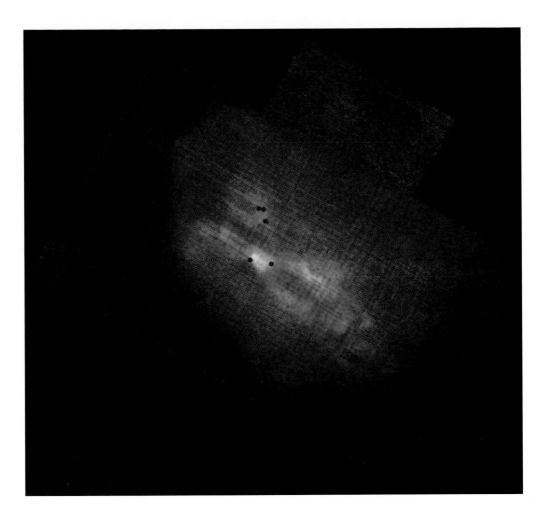

young stars with outflow activity. One is the infrared source that drives the HH 30 jet, seen in an optical image in Figure 6.11 obtained with the Hubble Space Telescope. The jet lies precisely in the plane of the sky, and the circumstellar disk of the source is perpendicular to the jet. The disk obscures the source and can be seen as a dark band between its outer parts that are illuminated by the central star.

Molecular outflows are ubiquitous. From statistical arguments it appears that most newborn stars, of both low and high masses, produce HH jets and molecular outflows. The most virulent outflow activity occurs while these stars are still embedded in their dark clouds and later tapers off.

As cloud material is swept up by outflows, cavities are created in the molecular cloud. Where many young stars are born, they create numerous cavities surrounded by expanding shells. As these shells collide and intersect, they churn the cloud. Large-scale organized motions degrade into smaller scale and lower amplitude turbulence. If enough energy is injected into the cloud, it can be completely disrupted. Figure 6.12 shows a cloud in the constellation Circinus, so full of cavities blown by outflows that it resembles a slab of Swiss cheese. By the time about 10 percent of a star forming dark cloud has turned into stars, the remaining 90 percent is often pushed away

Figure 6.10. The L1551 cloud in Taurus is one of the nearest star-forming regions and has spawned over a dozen young stars that are simultaneously driving jets into their surroundings. A major shocked outflow coincides with the blueshifted lobe in Figure 6.9. (J. Bally).

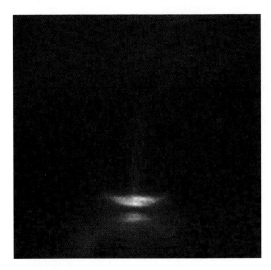

Figure 6.11. The HH 30 jet emanates from an edge-on disk around a young star located in the L1551 cloud. (NASA/STScI).

or destroyed by the action of its offspring.[3] Figure 6.13 shows a small cloud which has given birth to only a few stars. One is in the process of blowing a cavity, that is illuminated by the still embedded newborn star.

Properties of outflows

The interaction between fast Herbig-Haro jets and surrounding molecular material can sometimes be best traced by infrared observations. As fast gas slams into its molecular surroundings, the shock waves often have the right conditions to excite hydrogen molecules into emission. While fast shocks destroy molecules, in slower shocks molecules can survive. When shocks are bow-shaped, as is often the case in Herbig-Haro flows, the shock speed can take on a variety of values. For example, at the tip, gas comes in perpendicularly to the shock front, while in the wings it comes in at shallow angles. Thus at the tip of a fast bow shock, molecules are more likely to be dissociated and the resulting atoms may even be ionized. On the other hand, the much gentler shallow-angle entry of gas into the bow-shock wings may enable

Figure 6.12. A large dark cloud in the constellation Circinus shows a wealth of cavities and intricate structures caused by outflowing gas from newborn stars. (J. Bally & B. Reipurth, STScI/DSS).

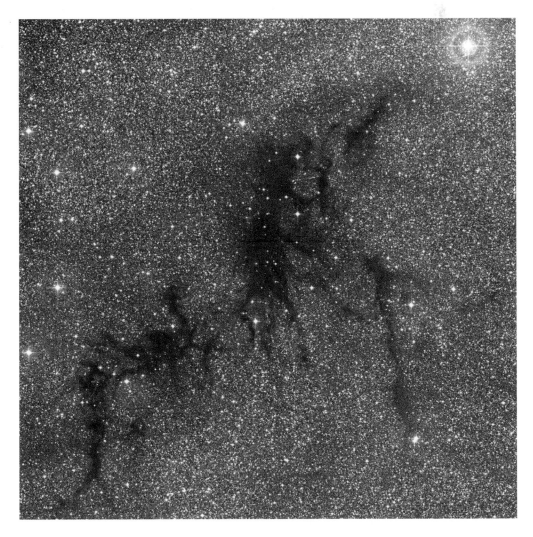

Figure 6.13. The small dark cloud BHR 71 contains a newborn star that is blowing a cavity in the side of the cloud. (Joao Alves, ESO).

molecules to survive. Therefore, while a bow-shock tip may predominantly glow in the light of various ions and atoms, the bow wings may be bright in the near-infrared light produced by molecular hydrogen. Most likely, CO is entrained and accelerated in the vicinity of glowing hydrogen molecules. This suggests that the carbon monoxide emission that traces molecular outflows is likely to be found in shells of swept-up gas surrounding the Herbig-Haro jets. Indeed, observations support this view.

Figure 6.14 shows HH 212, one of the finest bipolar jets known in the sky. The jet flows within a dark cloud so opaque that only one of the bow shocks can be weakly seen at optical wavelengths. The outflow structure is visible in the infrared due to the emission from molecular hydrogen molecules. The driving source is a deeply embedded Class 0 source.

While molecular hydrogen and the other tracers of jets and Herbig-Haro objects can only be seen as they are passing through a shock, CO does not require a shock to be visible. A radio telescope can see the emission from the CO molecule and measure its velocity along the line of sight under a wide set of conditions. The visible and near-infrared emission from an outflow traces the location of currently active shock waves. In contrast, the millimeter-wavelength CO emission traces *all* accelerated molecular gas. Thus, CO can trace the total mass, momentum, and size of an outflow, even if it does not contain any currently active shocks.

There is, however, a circumstance in which jets and non-molecular components of outflows can be seen in optical images without the presence of shocks. When a jet or outflow is lit from the outside by ultraviolet light of nearby massive stars, their lobes can become ionized, rendering them visible. Figure 6.15 shows a region with dark clouds inside the Trifid Nebula, which is illuminated by several very massive stars. The clouds develop a cometary shape as the powerful ultraviolet radiation sculpts the gas and dust in the region. From the head of the largest cloud, a finely collimated jet known as HH 399 bursts out, driven by a newborn star still embedded inside. The ultraviolet radiation from the nearby massive stars is ionizing the entire jet. Figure 6.16 shows another case of a young star, LL Ori, whose outflow is irradiated by ultraviolet radiation. In this case, the jet source is a visible star, and the two outflow lobes are bent sideways by a large-scale flow of hot, ionized gas (to be discussed further in Chapter 9). An irradiated bow shock is visible where this gas impacts the star and its outflowing gas.

There are several observed trends in the properties of jets and molecular outflows. The amount of energy and momentum in an outflow increases with increasing source luminosity and mass. More massive and luminous sources drive more powerful flows. Also, at least for the lower mass stars, the outflow properties vary with the evolutionary stage of the source star. The Class 0 protostars tend to drive very powerful flows. The rate at which energy is injected into the outflows of these young sources can be a substantial fraction of the total source luminosity. The flows from these extreme protostars tend to be very dense, have relatively low speeds, and are relatively

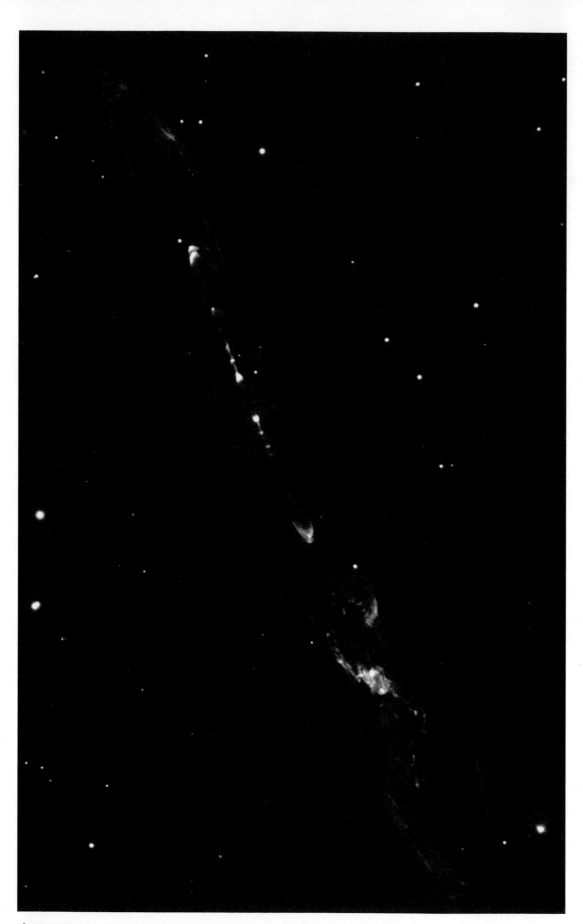

Figure 6.14. The embeddded bipolar jet HH 212 as seen at infrared wavelengths. A newborn star is hidden between the two lobes. (M. McCaughrean, ESO/VLT).

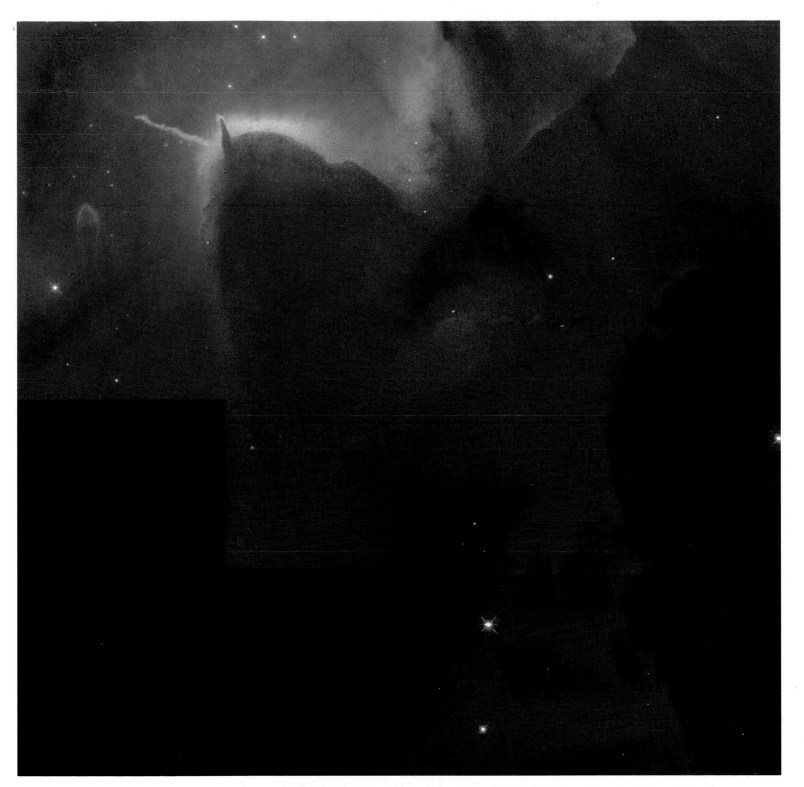

Figure 6.15. The HH 399 jet emerges from a dense globule into the powerful UV radiation from massive OB stars in the Trifid Nebula. (J.J. Hester, NASA/STScI).

Figure 6.16. The young star LL Ori produces a bipolar jet that is irradiated and bent by a large-scale flow from the left side of the image. (NASA/STScI).

compact. They emit predominantly in various molecular transitions, and their associated shocks are normally not visible optically, because the flows are still burrowing through the cloud surrounding the embedded driving sources. They can therefore be seen only at infrared and radio wavelengths.

As the source stars age, their flows become less dense, but develop faster speeds. Their emission tends to be dominated by atomic or ionic species rather than molecules. The flows often punch out from their molecular clouds, and their shocks become visible as Herbig-Haro objects. Such flows often reach parsec-scale dimensions. Where the jets emerge from their nascent cloud, they drag along gas swept up during their passage through the cloud core, so molecular outflows can in some cases extend beyond the boundaries of their host clouds.

As embedded young stars clear their environments and evolve into visible objects, their jets weaken further, and are often so weak as to be nearly undetectable except very close to the source. Though the most virulent outflows are driven by the youngest sources, faint jets and feeble outflows can sometimes be seen around stars that appear to be millions of years old.

Massive stars produce the most powerful outflows. While some young stars with luminosities of 10 000 times that of the Sun drive spectacular jets, others exhibit less collimated flows. A 100 000 solar luminosity massive star forming in the dense core immediately behind the Orion Nebula produced an explosive eruption about a thousand years ago that released more energy than the Sun radiates in ten million years. This powerful event is currently producing a very wide-angle bipolar outflow which is driving hundreds of individual shocks into the surrounding molecular cloud. We will discuss the highly energetic outflows from massive young stars in Chapter 9.

The youngest massive stars produce some exotic outflow phenomena that are only rarely found in lower mass protostars. Many massive star forming regions have been identified throughout the Galaxy by their intense radio wavelength *maser* emission. A variety of molecules, such as hydroxyl, water vapor, ammonia, methanol, and silicon monoxide (SiO), produce powerful emission at some radio frequencies when the environment is sufficiently dense and hot. For example in the hot molecular core known as Orion Molecular Cloud 1 (OMC1) located immediately behind the Orion Nebula, water vapor is heated to about 600 K. Here, densities are above a billion molecules per cubic centimeter and this gas is irradiated by infrared radiation emitted by nearby massive protostars. As a result, water and SiO molecules emit coherent radiation[4] that is millions of times brighter than the ordinary radio waves produced by warm molecules in the region. These masers are Nature's equivalent of lasers. The intense brightness of the various masers enables them to be studied with the extremely high angular resolution provided by intercontinental interferometers such as the *Very Long Baseline Array* (VLBA). The VLBA observations of the 1.3 cm water masers and 7 mm SiO masers provide an angular resolution of about 0.1 milli-arcseconds (a thousand times better than the Hubble Space Telescope) that corresponds to less than 0.05 AU in Orion. The masers trace clusters of knots in the ultra-dense parts of outflows and on the surfaces of disks embedded in exceptionally dense cores.

The ins and outs of young stars

The discovery that young stars in the process of assembling their mass also *lose* mass was unexpected. Astronomers have used a variety of observing techniques to study outflows from newborn stars, and we believe that we now understand at least the basic aspects of the phenomenon. But the fundamental question remains: why does this happen? If a young star is busy gaining mass, it seems counter-productive also to lose mass.

A variety of models have emerged in recent years to explain why and how young stars drive outflows. Many classes of astronomical objects produce winds, outflows, and collimated jets. Three types of driving mechanism have been proposed based on heat, radiation, and magnetic fields, respectively.

Above its visible surface, our Sun has an extremely hot atmosphere called the corona. This region can be seen during total solar eclipses as a silvery halo extending away from the Sun in all directions. The Sun's corona has temperatures that reach millions of degrees. Ions buzz around with speeds comparable to the gravitational escape speed from the solar surface, and as a result, the outer layers of the Sun's corona stream away into space as a wind. These ions and associated electrons sweep past the Earth with speeds ranging from 400 to 800 km/s. Although astronomers do not yet fully understand how the corona is heated, there are indications that solar flares and magnetic phenomena are responsible. Despite a lack of detailed understanding, we are confident that the solar wind is driven by such heating. During its 4.5 billion year history, the solar wind has removed much less than one percent of the Sun's mass.

As we will see in Chapter 9, massive stars produce winds which can be millions of times more powerful than the Sun's. Massive stars can lose ninety percent of their mass in their brief lifetimes of a few million years. The huge luminosities of massive stars drive these powerful winds, since the pressure of light can exceed the hold of gravity on the outer layers of such a star. Thus, massive stellar winds are examples of radiation driven winds.

As discussed above, young stars of all masses produce outflows. Unlike the solar wind, these flows are frequently collimated into bipolar jets. Despite their low luminosity and temperature, young low-mass stars can produce outflows that can be as powerful as the stellar winds powered by the most luminous stars. Thus, it is unlikely that either radiation pressure or heat can drive protostellar outflows. As a result, most astronomers have looked to magnetic fields as an explanation for such outflows.

Observations show that the entire plane of the Milky Way and its tenuous interstellar gas is permeated by a weak magnetic field about a million times weaker than the field found at the surface of the Earth. This field is strongly tied to the gas by the residual ions and electrons generated by the impact of energetic cosmic rays[5] and by ultraviolet light produced by hot stars. As atomic gas condenses into molecular clouds, the magnetic field is also compressed and amplified. Therefore, when a protostar and its disk form, they will be magnetized.

As a disk spins around its central protostar, entrained fields are stretched. Like strands of taffy, the field gets drawn out and amplified by shearing motions; consequently, circumstellar disks develop strong magnetic fields anchored to the disk. Models show that after a time, this disk field takes on a pinched hourglass configuration. Such disk magnetic fields can help accelerate outflows and collimate them into jets, as discussed below.

The energy released first by gravitational contraction, and later by thermonuclear fusion, drives convective motions in a star. Furthermore, as matter spirals in from the disk, young stars tend to spin up. Fast spin and convection are the ingredients of a magnetic dynamo. The weak fields inherited by a young star from birth become amplified by the dynamo. Observations show that young stars develop magnetic fields that can be hundreds of times stronger than the Sun's field and tens of thousands of times stronger than the Earth's. Models show that the most likely configuration of the stellar magnetic field will be similar to the Sun's, a so called "dipole field,"[6] more or less aligned with the rotation axis of the star.

Thus, protostars and their disks may have two types of magnetic field: a *disk* field and a *stellar* field.[7] There has been much debate about how magnetic fields launch jets, and interest has focused on the three general mechanisms discussed below.

First, the rapid collapse of a magnetized and spinning cloud core can overpressure the magnetic field that is dragged in. As first demonstrated by Uchida and Shibata in Japan in 1983, this powerful magnetic field can be wrapped up as the rotating protostar and disk form. Overstressed, the field recoils and launches a pulse of magnetic energy propagating above and below the disk. The magnetic wave accelerates gas above and below the disk plane and launches a powerful burst of outflowing matter. Models show that the magnetic field can collimate the outburst into a pair of oppositely moving jets. However, this mechanism produces a powerful transient eruption and not a steady outflow, so it is unlikely to be responsible for the long-lasting Herbig-Haro jets.

Second, as the disk settles into a state of equilibrium, its pinched hour-glass magnetic field geometry may lead to a less powerful but steady wind. Charged particles can only move along magnetic field lines and not across them. Above and below the disk plane, the magnetic field rotates with the disk. If the field lines open away from the central star, as implied by the pinched hourglass shape, particles will be flung out and away from the disk. Like a slingshot, ions and electrons are accelerated by the outward facing lines of magnetic force, and these charged particles drag along neutral hydrogen and helium. The result is a steady bipolar wind. It is thought that far above and below the disk, the magnetic field will eventually become parallel to the rotation axis of the star. As the wind enters this region, it becomes collimated into a jet. Theoretical models indicate that the magnetic fields play a central role in removing angular momentum from the disk. Thus, magnetic fields in circumstellar disks may not only produce collimated jets and outflows, they may also enable the removal of spin energy from the disk.

Third, the dynamo-generated stellar magnetic field may also produce an outflow. One of the principal researchers in this field, Frank Shu, has developed what is perhaps the most complete model of jet formation so far. In this picture, the stellar magnetic field interacts with the magnetic field of the inner disk to produce a gap between the star and the disk. Because of the strong magnetic linkage between star and disk, the rotation of the star becomes locked to the orbital speed of matter at the inner edge of the disk. If the star and inner disk did not rotate at the same speed, the magnetic fields would start to wind up and force the star and inner disk edge back into co-rotation. As material spirals inwards through the disk and reaches the inner disk edge, it divides into two streams. A high angular momentum stream is flung away along the rotating magnetic field lines, and a low angular momentum stream falls onto the star and helps build its mass. The gas flung away along the magnetic field lines rising above the disk will form a highly collimated Herbig-Haro jet as observed from many of the youngest stars.

The last two models produce continuous streams of outflowing gas. But Herbig-Haro jets are irregular and contain a multitude of knots and larger bow shocks. A mechanism must exist to modulate the outflow. Instabilities in a circumstellar disk may affect the outflow velocity, producing a pulsed jet. Alternatively, if the source is a close binary in an eccentric orbit, every time the two components approach each other, their disks become perturbed. Such perturbations produce enhanced accretion onto the star and at the same time enhances mass loss, so the periodic close passage of a companion can result in periodic bursts of outflow activity.

Observational techniques are not yet good enough to directly observe the regions where jets are launched and collimated. In the absence of strong empirical guidance, astronomers actively discuss the finer points of the three models presented above. But while there may be debate about precisely how jets are generated, there is general consensus that mass loss from young stars is a consequence of accretion from a rotating system. By flinging out a fraction of the accreting matter, jets can carry away most of the angular momentum, leaving the remainder of the gas to fall onto the young star. Jets are one of Nature's ways of solving the problem of how to build a star from material with angular momentum.

7 Towards adulthood

Protostars are found inside dark clouds as infrared sources, while pre-main-sequence stars are usually visible stars scattered near molecular clouds. Observations show that most stars become visible at visual wavelengths within about 100 000 years of their formation. But how do stars emerge from their parent cores and host molecular clouds?

Most stars are born in large clusters, as will be discussed in chapters 8 and 9. Often such clusters contain a few massive stars, whose light and winds have devastating effects on the surrounding clouds and thus aid in extricating their neighboring lower-mass stars. However, some stars are born in smaller, loose associations where they do not have this kind of help, and they must resort to other processes.

As we saw in the previous chapter, forming stars produce high-velocity jets and winds which significantly alter the clouds out of which they formed. These outflows are so powerful that they can destroy a cloud core and thus liberate an embedded young star.

In principle, there is another way for a young star to get rid of the cloud it is born from: it can simply move away from it. This can happen when the star is born in a small cluster of young stars that interact dynamically. As we saw in Chapter 5, stars in small groups continuously move among each other, and on occasion one can be ejected with enough speed to escape the gravitational field of the cluster and cloud. How often this happens is unclear, but some researchers believe that a significant number of stars escape their birth sites in this manner.

Properties of young stars

Sixty years ago, Alfred Joy noted a dozen stars around the sky that varied irregularly in brightness on time-scales of days, and which were all associated with dark or bright nebulae. Joy called these stars *T Tauri variables*, after T Tauri, one of the brightest members of this group. At the time it was not known that these stars are young. Today T Tauri stars are seen as infant low-mass stars which will eventually mature into stars similar to our Sun.

The spectra of T Tauri stars show certain peculiarities that distinguish them from other, older stars of low mass. In particular, T Tauri stars show bright emission lines such as the Hα line of hydrogen.[1] In fact, so bright is the Hα line in many T Tauri stars, that it has become the main characteristic used to discover new members of the class within regions of recent starbirth. A small prism in the light path of a telescope in front of a camera will spread the light of all the stars in the field into small spectra, quickly revealing stars with the Hα line in emission. Such Hα emission line surveys have been carried out for all the nearest star forming regions, and many thousands of young stars near the Sun have been identified.

Another signature of stellar youth is a dark absorption line, due to the element lithium, in the red part of the spectra of T Tauri stars. This line has particular significance, because lithium is destroyed in nuclear processes at a relatively low temperature. Thus, in main-sequence stars, lithium has been destroyed, as surface layers are constantly dragged down to the hot core where lithium is burned. T Tauri stars, on the other hand, are so young that nuclear processes have not yet started. Lithium is therefore visible in their spectra as a signpost of their youth. Looking for lithium is a common way of establishing the youth of low-mass stars.

T Tauri stars display other oddities in their spectra. In the blue and ultraviolet spectral regions they are brighter than normal stars of the same spectral type. And in the infrared, T Tauri stars also tend to be much brighter because of emission from circumstellar disks. These characteristics offer important clues to how T Tauri stars and their disks evolve. The infrared excess in T Tauri stars has been an especially useful characteristic when studying young stars that are too shrouded to be seen in visible light. Examining the infrared colors of stars can help identify young embedded stars.

T Tauri stars tend to have strong magnetic fields, huge dark star-spots, and brilliant flares millions of times stronger than those on the Sun. While still accreting from their disks, T Tauri stars emit orders of magnitude more UV and X-ray radiation than the Sun. Some even emit radio waves. Thus, erratic variability, flaring, X-ray, UV, and radio emission provide observers with unique observable characteristics that can distinguish young stars from more mature main sequence stars.

As they emerge from their birthsites, T Tauri stars are considerably cooler and more luminous than main-sequence stars of the same mass. Solar mass stars are born over-luminous by a factor of 100 compared with mature stars like the Sun. These stars undergo two distinct types of evolution. First, T Tauri stars shrink and become less luminous at a more or less *constant surface temperature* and color. This is the Hayashi track already mentioned in Chapter 4. Subsequently, the T Tauri stars become hotter at *constant luminosity*. In this phase, the star has its final luminosity, but is much cooler than a main sequence star of the same mass. As the outer layers of the T Tauri star continue to shrink, the star becomes hotter and bluer, until it reaches the main sequence and starts burning hydrogen in its interior. This whole evolution may take about 10 million years for stars like the Sun. Very low-mass red dwarf stars are more than a thousand times more luminous than main-sequence stars of the same mass. The lowest mass stars take more than one hundred million years to reach maturity. On the other hand, stars that are five or ten times as massive as the Sun reach the main sequence in less than a million years.

Circumstellar disks

There have been numerous attempts over the years to explain the unique characteristics of T Tauri stars, but until recently all have turned out to be unsatisfactory. The key to understanding T Tauri stars is the recognition that they are surrounded by disks. The initial evidence for disks was circumstantial. Among other things, it was noted that when spectra of T Tauri stars show signs of high velocity shocks associated with outflows, the emission is usually blue-shifted, that is turned towards the observer, whereas red-shifted shocks are rare. These shocks are likely to be generated by material moving equally towards and away from the observer, and therefore such a preference should not exist. But if T Tauri stars are surrounded by disks, the shocks moving away from the observer will be partly hidden, while the approaching shocks will remain unobstructed.

After the Hubble Space Telescope was launched, astronomers had the tools required to directly image circumstellar disks around T Tauri stars. Examples of dark dust disks are seen in the figures of Chapters 4 and 12. Many examples of such disks are now known, confirming their ubiquity.

Starlight heats the inner portions of circumstellar disks, and absorbed light is re-emitted at infrared wavelengths. The hot inner edge of a disk will glow in the near-infrared, and the cooler outer disk will emit in the far-infrared portion of the spectrum. Still cooler dust further out in the disk emits in the sub-millimeter and millimeter wavelength regimes. Thus, disks emit radiation that can be detected and studied in detail with infrared telescopes and millimeter interferometers. The short wavelengths probe the inner disk and longer wavelengths probe the outer disk regions. Such data provide measures of disk masses and compositions. Disks typically have masses between one-hundredth and one-tenth of the mass of the Sun, consistent with the amount of material required to form a planetary system. Furthermore, observations show that disks rotate in response to the gravity of their central stars. The usual tracers of gas in molecular clouds such as CO tend to be poor tracers of disks, because these species are severely depleted as they freeze onto grains. However, molecules trapped by ices on grain surfaces, and the materials from which the grains are made, can be observed at infrared wavelengths. Infrared observations provide data on the composition and physical properties of disks. It is likely that the disks around young stars represent a present-day version of the material that surrounded the early Sun and out of which our planetary system formed.

Disks usually have gaps between the star and an inner edge at 5 to 10 stellar radii. These gaps can be carved by intense magnetic fields (Chapter 6),

Figure 7.1. An artist's impression of a young star surrounded by an accretion disk. The strong stellar magnetosphere disrupts the inner disk and funnels gas to the star. A bipolar wind escapes orthogonally to the disk. (P. Hartigan).

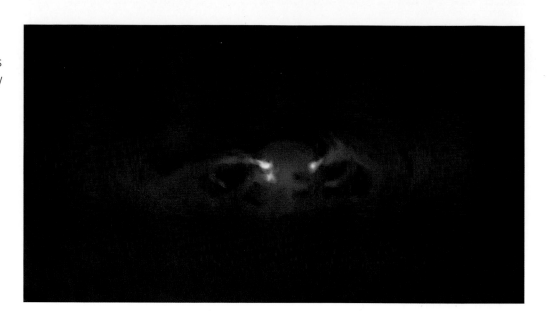

radiation fields (Chapter 12), or by forming proto-planetary condensations (Chapter 10). Although we do not have the resolution to observe such gaps directly, calculations predict that if they were not there, T Tauri stars would emit a lot more light in the infrared spectral region than what is observed.

Young stars are expected to rotate fast around their axes, because after birth a star slowly contracts until it reaches adulthood. Like the ice-skater of Chapter 4, young stars should spin up as they shrink due to conservation of angular momentum. However, contrary to expectations T Tauri stars are often surprisingly slow rotators. It is likely that disks are responsible for braking stellar rotation.

T Tauri stars have strong magnetic fields. As already discussed in Chapter 6, disks are threaded by magnetic fields left over from the clouds from which they formed and magnified by the compression of collapse. The two magnetic field systems form a bridge between star and disk, and the stellar spin thus becomes locked to the orbital speed of the disk inner edge.

Gas in a T Tauri disk orbits the young star as planets move around the Sun following laws established four hundred years ago by Johannes Kepler. This Keplerian rotation dictates that gas closer to the center moves faster than gas farther away.[2] But gas molecules constantly collide and exchange momentum, and as a result of complex processes most of the gas will move slowly inwards, while angular momentum is transferred outwards.[3] Thus it is possible for gas from the infalling envelope that lands on the disk to spiral through the disk towards the star.

Eventually the gas comes to the inner edge of the disk where it encounters the gap between star and disk. There it latches on to the magnetic field lines that link the disk with the star. Using the magnetic field as a bridge, some of the gas travels through the gap in so-called funnel flows, eventually crashing onto the stellar surface. In this manner the star is able to grow from the material in the disk. Figure 7.1 shows an artist's impression of what a

T Tauri star may look like as it is fed by several funnel flows from its disk. As discussed in Chapter 6, the conservation of angular momentum requires that some gas is also expelled to large distances in the form of a bipolar jet. Estimates vary, but probably between 10 and 30 percent of the gas moving through the disk will be ejected.

As mentioned earlier, T Tauri stars exhibit excesses of light in the blue spectral region as well as at near-infrared wavelengths. The near-infrared excess is produced by the heating of the disk both by the stellar radiation and by internal friction. On the other hand, the excess of blue light is radiation from hot spots on the star where material in the funnel flows crashes onto the stellar surface, releasing energy and locally heating the gas to about 20 000 degrees. This hot plasma radiates mostly in the blue and ultraviolet spectral regions.

The chemistry of the disk is of particular importance, because planets will eventually form out of this material. Since a disk consists of gas and dust from the molecular cloud out of which the associated star was born, we know the abundances of the elements incorporated into the disk, but not the chemical compounds formed by these elements. The chemistry of the disk depends sensitively on the temperatures and densities and therefore varies significantly throughout the disk and also with time as the disk evolves. In Chapter 10 we discuss the processes that lead to the formation of planets from circumstellar disks.

Eruptions

In 1936 a faint star in the constellation Orion, called FU Orionis, erupted in a major explosion which led to a hundred-fold increase in its brightness in less than a year. Thirty four years later, in 1970, another eruption took place on a faint star in the constellation Cygnus. Both stars then began a very slow decline in brightness, but at different rates. Whereas the Cygnus object is today almost back to its earlier faint state, FU Orionis is only slightly fainter than at maximum. In a pioneering study, George Herbig noted the similarity between the two events, pointed out that both progenitor stars were likely to be young, and suggested that these outbursts, now known as *FUor eruptions*, were an important element in the early evolution of young stars. A handful of objects have subsequently been found that share the characteristics of FUors, even though their eruptions were not observed in all cases.

There are many types of outburst phenomena among evolved stars, such as nova and supernova explosions. So why did Herbig conclude that FUors are pre-main-sequence objects? First, FUors are associated with dark clouds, and the velocities of star and cloud are always similar. Second, FUors illuminate large bright nebulae, as light from the star is reflected by a nearby dust cloud.

FUors do not look like T Tauri stars. Their appearance depends on the wavelength range observed. In the blue spectral region they resemble hot stars, in the yellow they mimic the Sun, and in the red and infrared they look like very cool stars. They do not have the Hα line prominently in emission like T Tauri stars. And their spectra reveal prodigious mass loss in very powerful winds.

In a theoretical analysis, Lee Hartmann and Scott Kenyon proposed an explanation: FUors could be T Tauri stars whose disks had begun to transport a thousand times more material than normal. Because different annuli of a disk rotate at different velocities, they rub against each other and the friction heats the disk material. In T Tauri stars such disk heating is, in most cases, limited to emission in the near-infrared spectral range. But if a disk has to accommodate a thousand-fold increase in the rate at which gas spirals in, the disk rapidly heats, swells, and brightens. The disk could become so large that it completely swamps the star. The light we see from a FUor may therefore not be stellar light, but the emission from a glowing disk heated by the rapid release of gravitational potential energy as matter spirals ever deeper into the gravity well of the central star. While this hypothesis successfully explains many of the characteristics of FUors, a number of puzzling details remain, which astronomers are actively trying to understand.

Based on statistical arguments, Herbig demonstrated that FUor outbursts must be repetitive. Any given young star may undergo a dozen or more such eruptions before settling on the main sequence. Each time a young star erupts, it accretes a great amount of mass. It is estimated that for each FUor outburst the underlying star gains about one-hundredth of the mass of the Sun, or about 10 times the mass of Jupiter. If a star experiences ten FUor outbursts, it gains one-tenth of a solar mass. Thus, a large fraction of the mass a star accumulates during the T Tauri phase may accrete in a few massive spurts.

FUor outbursts are recognized as rare but important phenomena in the life of a young low-mass star, and much effort has been spent trying to understand why a young star suddenly flares so violently. There are two hypotheses. One model is based on the inherent instability of hydrogen in disks. As matter spirals toward the central star, friction injects heat into the disk. Frictional heating tends to increase towards the central star. But the dissipation of orbital energy and angular momentum, and therefore the rate at which gas is heated, depends critically on the phase of the gas. Dissipation and heating is low in cool, mostly neutral gas; it is high in hot, mostly ionized gas. If a disk contains enough mass, it can make an abrupt transition from a cool, low dissipation disk to a hot, highly dissipative one. This instability starts in the densest portion of the disk close to the star where dissipation is at a maximum. If the temperature in this region reaches the critical level where hydrogen becomes ionized, the dissipation rate will increase dramatically. Increased dissipation heats and ionizes more hydrogen, leading to even more dissipation. The disk then suffers cataclysmic runaway heating. As more of the disk becomes ionized, the accretion rate onto the central star increases. The disk becomes self-luminous and may outshine the central star a hundredfold or more. Finally, rapid accretion onto the central star drains

Figure 7.2. When a circumstellar disk of a young star in a binary system is penetrated by a companion star it leads to severe disk disturbance and truncation. (A. Whitworth & H. Boffin).

the inner disk and the eruption subsides. Subsequently, the disk is gradually rebuilt by infall from the surrounding envelope until another runaway heating event drains it again.

Another possibility is that FUors are newborn binary systems. The presence of large amounts of gas in the form of circumstellar and circumbinary disks leads the two stars to gradually spiral towards each other. In the process, their disks interact vigorously (Figure 7.2), and this may also lead to FUor outbursts.

The last major FUor outburst occurred a third of a century ago. Astronomers have been waiting for another such event, so that modern instruments can be brought to bear on this phenomenon. It was therefore with great excitement that an amateur astronomer, Jay McNeil, from Kentucky, USA, in January 2004 announced that a new nebula had appeared in an otherwise dark region of the L1630 molecular cloud in Orion. Figure 7.3 shows an image of the L1630 cloud taken by Jay McNeil. The large nebula is known as Messier 78, a reflection nebula illuminated by a more massive young star. McNeil's nebula is seen further down along the dark lane as a small cometary shaped nebula. Figure 7.4 shows a higher resolution image of McNeil's nebula taken with the Gemini 8 m telescope in Hawaii. The partly obscured star at the base of the nebula has undergone a major accretion event and as a result has brightened dramatically, in the process illuminating the large outflow cavity just north of it.

Figure 7.3. The Messier 78 region in the L1630 cloud in Orion. McNeil's nebula is seen further down along the dark cloud. (J.W. McNeil).

Farewell to disks

In chapter 3 we discussed how protostars can be divided into Class 0 and Class I objects depending on the wavelength range in which the object emits most of its light. Class 0 objects are so deeply embedded in their natal material that they can only be detected at sub-millimeter wavelengths as extremely cold objects. Class I objects are still embedded in massive circumstellar material, but are more evolved and can be best detected at infrared wavelengths.

As newborn stars emerge from their placental material, they become optically visible, at which stage we call them T Tauri stars or, in continuation of the above scheme, Class II sources. In these objects, starlight dominates the disk emission, so Class II objects are usually bright at visual wavelengths. Because they are surrounded by large circumstellar disks, they are also bright at infrared wavelengths.

As young stars evolve, the infalling envelopes that fed the stars and their disks disappear. At this stage a disk can no longer get replenished as it processes material and feeds the central star. Disks are therefore expected to

Figure 7.4. McNeil's nebula as seen with the 8 m Gemini telescope in Hawaii. The nebula is an outflow cavity illuminated by the erupting star at its base. (B. Reipurth & C. Aspin, Gemini/AURA).

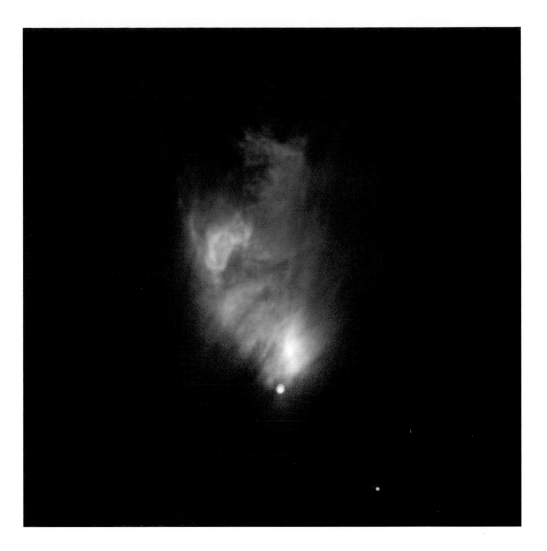

diminish in mass as the stars age and mature. Since most characteristics of T Tauri stars or Class II sources are due to disks, it follows that we should be able to find a population of more evolved young low-mass stars with few or no signatures of disks.

Indeed such stars exist and are known as *weak-line T Tauri stars* or Class III objects. As the name suggests, these stars are characterized by very weak Hα emission. They also have virtually no infrared emission in excess of that expected from a star of the same spectral type. This does not mean that these objects are completely devoid of circumstellar matter, rather the material that is left over in their circumstellar disks may have begun to coagulate into larger bodies, which are building blocks for the construction of planets. If much of a disk's mass is incorporated into such large bodies, it becomes invisible at most wavelengths. Class III stars represent the active phase of planet building, as discussed in detail in Chapter 10.

How long do T Tauri stars maintain their characteristics? How long do disks survive? Statistical studies aimed at answering these questions suffer from the complication that, as they age, T Tauri stars do not seem to move gently from the Class II stage into the Class III stage. Rather it seems that

Figure 7.5. The Sun with several large sunspot groups as it appeared on October 29, 2003. (Meese Solar Observatory).

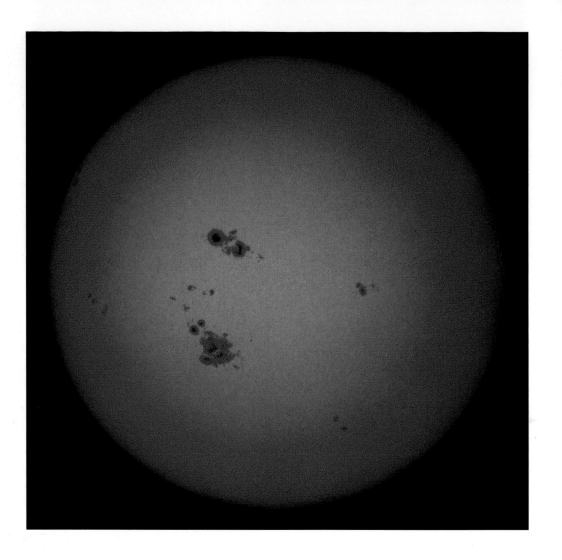

a star may for some time oscillate between the two stages, with a T Tauri star going through periods of dormancy with little activity, and Class III objects occasionally waking up to periods of increased activity. Nonetheless, we can get an idea about disk lifetimes from the fraction of young stars in clusters of different ages that still have a measurable infrared excess emission produced by their disks. Such observations show that disks start to disappear in a few million years, and after 6–10 million years virtually no stars with infrared excesses are left. At that age, the material in a disk has either been swallowed by the star, blown away, or has condensed into grains, gravel, and boulders in preparation for the formation of planets.

Adolescence: spots, flares, and X-rays

T Tauri stars are variable and flicker irregularly on time-scales of days. This variability is caused by material that rains from the disk onto the stellar surface. As disks are gradually depleted, young stars diminish their irregular variability. When stars reach the Class III stage, a low-level cyclical variability becomes apparent as the irregular variability subsides.

Figure 7.6. A detailed sunspot picture taken with the Swedish Solar Telescope at La Palma. The granulated surface is due to convective motions in the Sun. (Inst. for Solar Physics).

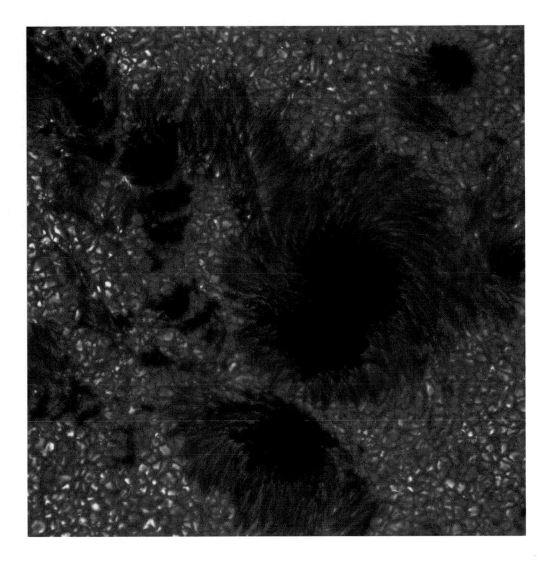

The cyclical variability may be produced by giant star-spots similar to those seen at the Sun, but much larger. Figure 7.5 shows an image of the Sun with a group of dark spots. Such spots are not really dark, they appear dark only in contrast to their surroundings, because they are a few thousand degrees cooler than the rest of the solar surface. Sunspots form where strong magnetic fields well up to the surface, causing the temperature to drop locally (Figure 7.6). As a T Tauri star rotates, the presence of a major spot group covering a sizable fraction of the stellar surface will modulate the light from the star with its rotation period. By measuring the periodic variations in a star's light, rotation periods of numerous young stars have been determined.

Young stars, both Class II but especially Class III objects, occasionally display brief outbursts, during which the brightness of the star rapidly increases several-fold, followed by a slower decay over several hours. Little is known about such *flares*, mostly for lack of facilities to monitor young stars for long periods of time. But from serendipitous observations, it appears that such flares represent magnetic activity on the stellar surfaces.

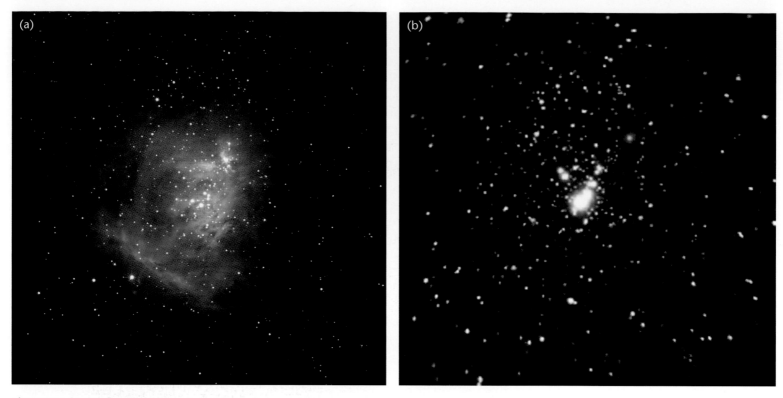

Figure 7.7. (a) An infrared and (b) an X-ray image of the Orion Nebula Cluster. (David Thompson/Palomar, NASA/CXO).

Magnificent tools for studying magnetic fields on young stars have become available in recent years. In the last decade it has been established that young stars are copious emitters of X-rays. Since X-rays are absorbed in the Earth's atmosphere, such observations require dedicated space observatories, like *Chandra* and *XMM-Newton* (see Chapter 2). X-rays have about the same ability to penetrate dusty interstellar clouds as near-infrared light, and the two observing techniques detect about the same number of young stars. Figure 7.7 shows two images of the same part of the Orion Nebula Cluster; an infrared image (observed at 2 microns), and one from the Chandra X-ray satellite. Since X-rays cannot be focused with normal optical systems, rather unusual telescopes are required to form an X-ray image, and as a consequence the outer regions of the X-ray image are not as well focused as the center. If we disregard this effect, the infrared and X-ray images show virtually the same population of young stars.

To understand X-ray emission from young stars, we can get useful guidance from similar studies of our own Sun. Because of its proximity, we can get gorgeous images of the solar surface at very short wavelengths showing large luminous clouds that protrude from and envelop the solar surface (Figure 7.8). These huge billowing plasma clouds of highly ionized gas at temperatures of about 10 million degrees are confined by magnetic loops anchored in the Sun. Long-term X-ray monitoring of the solar surface shows that all such structures continuously evolve, with occasional bright flares lasting a few hours. X-ray flares trace the reconnection of magnetic field

Figure 7.8. An extreme ultraviolet image of the entire solar disk shows numerous areas of very hot gas associated with strong magnetic fields. (NASA/ESA/SOHO).

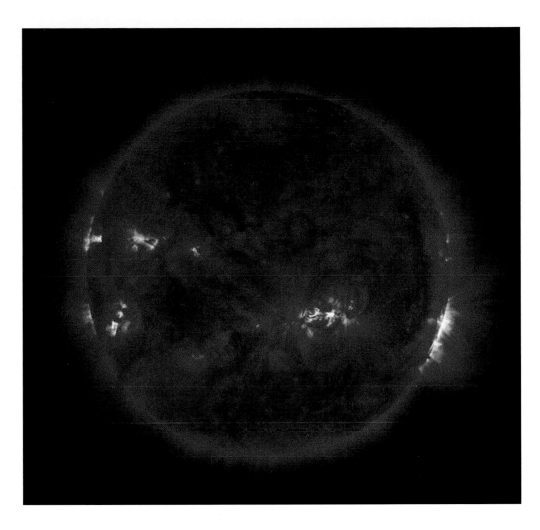

lines generated by dynamos under the solar surface driven by large-scale convective motions.

T Tauri stars emit more than a thousand times more X-ray radiation than the Sun when it is active, with Class III sources on average being more X-ray luminous than Class II sources. Giant X-ray flares are common among these young stars and are believed to be produced by hot plasma with temperatures up to 100 million degrees in scaled-up versions of the solar flares. The magnetic loops may be gigantic, with sizes as large as the stars.

Since most T Tauri stars have circumstellar disks, and since inner edges of the disks are locked to the stars via magnetic fields, it is possible that at least some observed X-ray activity may be driven by interactions between stars and their disks. Some giant flares observed in deeply embedded outflow sources may represent outbursts when magnetic fields between star and disk reconnect and release huge amounts of energy. These issues are currently subject to considerable research.

The intense X-ray activity of Class III sources occurs at the same time that planets are thought to form. As X-rays are released, large swarms of particles are accelerated to relativistic speeds and bathe the rocks and solid bodies that will form new planets. Much effort has been expended analyzing meteorites (see Chapter 11) for effects of X-ray irradiation and collision with energetic particles. Such measurements indicate that our Sun apparently went through very active phases similar to what is now observed in Class II and Class III sources.

During the Class III phase, young stars calm down from their adolescent antics. They continue to slowly contract, and eventually, after 10 or 100 million years (depending on mass), the temperature and pressure becomes high enough in their interiors to settle into stable hydrogen burning. At this point, young stars begin their long adult lives, which for low-mass stars like our Sun will last about 10 billion years.

Herbig Ae/Be stars

Stars with masses in the range of 2 to about 5 solar masses form and evolve much faster than lower mass stars. Owing to their greater mass, these stars have between 10 to 1000 times the luminosity of the Sun. Nevertheless, they do not have enough mass to produce the hard ultraviolet radiation that can ionize hydrogen (we will discuss the birth and impact of stars more massive than eight solar masses in Chapter 9). These stars are appropriately called intermediate mass objects. Though they are born in a manner similar to low-mass stars, they evolve onto the main sequence in only about a million years. Unlike their lower-mass siblings, intermediate-mass stars do not evolve through a prolonged T Tauri phase. Many of these stars are of spectral type B or A with surface temperatures quite a bit hotter than the Sun.[4] They frequently exhibit bright emission in the light of various elements such as hydrogen. George Herbig was the first astronomer to realize that hot stars embedded in reflection nebulae with bright emission lines may represent the more massive analogs of T Tauri stars. Thus, they have been dubbed Herbig AeBe stars (the A and B stand for their spectral types while the lowercase e stands for "emission line") or HAeBe stars for short.

HAeBe stars are often found within reflection nebulae, demonstrating their close association with molecular clouds. Figure 7.9 shows an image of a star-forming molecular cloud in the southern constellation Corona Australis. The two largest reflection nebulae in the northern part of the cloud are illuminated by the luminous HAeBe stars, R CrA and TY CrA. Like many other HAeBe stars, they are surrounded by dozens of nearby T Tauri stars with which they share a common motion through space. This association makes it very likely that the HAeBe stars formed from the same cloud and at roughly the same time as the T Tauri stars.

Relatively few HAeBe stars show clear signatures of circumstellar disks. Nevertheless, infrared and sub-millimeter emission from a few indicate the presence of disks. Furthermore, some HAeBe stars drive jets. Interestingly, most intermediate-mass stars rotate much faster than low-mass stars. One interpretation of these observations is that the intense radiation fields of the HAeBe stars leads to such rapid disk dissipation that by the time the

Figure 7.9. A large dark cloud in the constellation Corona Australis with several bright young HAeBe stars enveloped in reflection nebulae. (ESO).

stars become visible, they have lost their disks. Just as FU Ori eruptions may be driven by the self-ionization and fast drainage of a disk onto the central star, so the heating of the disk by intense stellar ultraviolet radiation may promote rapid accretion and disk dissipation.

Models of stellar structure indicate another difference between T Tauri stars and HAeBe stars. The lowest mass stars transport the energy generated in their cores to their surfaces by large-scale convective mass motions. In stars like the Sun, the cores and intermediate layers are so hot that energy can be transported just by radiation, and convection occurs only in the outer layers.[5] As the mass of a star increases, the thickness of the convective layer diminishes, and eventually disappears. Astronomers suspect that convection drives magnetic field generation in low-mass stars. Thus, moderate-mass stars, which have very thin or no outer convective layers, may not support stellar magnetic fields. Without such fields, their inner accretion disks may not be disrupted and may not develop central holes. If so, matter from the disk is more likely to accrete directly onto the stellar equator, and spin up the star.

The IRAS survey and other infrared searches have revealed a number of protostars embedded within molecular clouds with luminosities of 100 to over 1000 solar luminosities. These stars are likely to be the progenitors of HAeBe stars. They tend to drive very powerful outflows, and are surrounded by massive disks. The main difference between these objects and the lower mass protostars is that by the time they emerge from their birth sites, they are already mature stars that have lost their outflows and have shed their disks. We will discuss the birth and impact of even more massive stars in Chapter 9.

8 The social life of stars: stellar groups

Associations: a loose brotherhood of stars

The nearest star forming regions are found in loose dark cloud complexes, such as the one in the constellation Taurus, which contains a few hundred low-mass stars formed within the past few million years. Such groupings are called *T associations*, after their namesake, the young star T Tauri.

The vast majority of low-mass stars are born from giant molecular clouds, which spawn many thousands of stars, and not in T associations. In addition to the low-mass stars, these giant molecular clouds also give birth to luminous, massive stars (Figure 8.1). Although such high-mass stars are relatively rare, they act as beacons of recent star formation, because they are blazing with a brilliance of up to almost a million Suns. These massive denizens of our Galaxy are called O and B stars after their designations in the stellar classification scheme invented during the early part of the twentieth century. Hence, the loose clustering of bright massive stars, along with their lower mass brethren, are known as *OB associations*. An example is the group of young OB and T Tauri stars which illuminates the nebula IC 1805, located at a distance of 7500 light years (Figure 8.2).

Anyone who has spent time under the clear southern skies in May or June has seen the splendor of the hundreds of bright stars sprinkled overhead along the Milky Way arching from the Scorpius/Sagittarius border to the rich star fields of Centaurus and the Southern Cross. These naked eye stars are the most luminous members of the nearest OB association, known as the Scorpius/Centaurus OB Association, less than 400 light years from us.

On the opposite side of the sky lies Orion, about 1500 light years away. Young massive stars dominate this constellation. Additionally, the giant molecular clouds in Orion have produced many tens of thousands of low-mass stars since they started forming stars about ten million years ago.

During the 1940s, the Dutch astronomer Adrian Blaauw noticed that Orion's clouds gave birth to four distinct sub-groups of stars during periods of unusual star formation activity. Blaauw realized that these young stars are no longer at their place of birth. While the younger groups are very compact, the older ones are more spread out. This could only mean that after their birth, young stars begin to migrate and disperse.

Figure 8.1. A small newborn cluster of heavily reddened stars is seen in this infrared image that probes into dense dark clouds in the NGC 2024 region of Orion. (David Thompson/Palomar).

Figure 8.2. The dramatic effect of an OB association on its surroundings is illustrated by this image of IC1805. A piece of the left-over parent molecular is being sculpted by the intense UV radiation of massive stars. (J.-C. Cuillandre/CFHT).

Into the void

The loose groupings of stars with a common protostellar heritage within both T and OB associations are not permanent. Member stars slowly drift apart as can be demonstrated by careful study of their motions. For example, the bright star Betelgeuse in Orion has moved over 150 light years from its birthplace during its 10 million year life. Measurements of other bright members of the Orion OB association show that most are moving away from their birth sites with typical speeds of a few kilometers per second. This implies the stellar groupings in Orion are transient.

Why do young stars drift away from their nests? The reason has to do with the mutual gravitational attraction between the stars and their associated clouds. Before giving birth to stars, molecular clouds are bound by their own gravity. Therefore, the young stars that form from such clouds are bound to them and inherit their small random motions.[1] But as we have seen, the action of outflows and intense starlight disperses the cloud. As the cloud is gradually blown away, the hold which the cloud exerts on its young stars weakens. Eventually, the stars are left on their own. In most cases, the mutual attraction between the stars is insufficient to bind them to the parent cluster, and they start to drift away.

The dark cloud is the gravitational glue that holds young stellar groups together. When most of the gas is dispersed, this glue is lost and the stellar group expands and dissolves. In the end, its young members join the general population of *field* stars of the Galaxy.

The birth of clusters: ties that bind

Some star-forming regions leave behind spectacular clusters of stars so numerous that they remain bound to each other by their mutual gravity even after cloud dispersal. Most common are the so-called *open clusters* which contain anywhere from dozens to thousands of stars. A beautiful example is the Pleiades cluster, a group of about 500 stars in the constellation Taurus which can be seen with the naked eye in the northern winter sky (Figure 8.3). This cluster has an age of about 80 million years and the placental cloud was dispersed a long time ago. The nebulous sheets and filaments surrounding the Pleiades stars trace a chance encounter of the cluster with a diffuse cloud. Such encounters are relatively rare and, when they occur, give researchers an opportunity to study the structure and properties of diffuse non-star forming gas and dust (Figure 8.4). Another example of an open cluster is Messier 11, a grouping of more than 3000 stars born over 200 million years ago (Figure 8.5).

The numbers and ages of open clusters in our Milky Way indicate that they originated in very massive OB associations that formed in exceptionally rare but powerful bursts of star formation. The scarcity of open clusters compared to OB associations suggests that only few of the currently forming OB associations in the neighborhood of the Sun are likely to leave behind open clusters. Astronomers estimate that in our portion of the Galaxy, four bound star clusters form every 10 million years out of each 3000 by 3000 light years region projected on the plane of the Milky Way. Even though most stars form in OB associations, most will not be bound to a cluster.

The birth of open star clusters, where stars remain gravitationally bound after dispersal of their parental clouds, requires that a large fraction of the initial cloud mass be converted into stars. This again requires a high star formation efficiency.[2] The cloud must have sufficiently high density and high pressure so that the outflows from low-mass stars and the UV radiation

Figure 8.3. The Pleiades is a well-known cluster in Taurus which at present happens to move through diffuse material that it illuminates. (Johannes Schedler).

from massive ones do not end star formation. Star formation must be allowed to continue until at least half of the initial cloud mass is converted into stars before star formation is brought to a halt. Then, when the cloud is dispersed, there will be sufficient mass left in stars alone to gravitationally bind the cluster. Though the cluster may expand somewhat after birth, it may survive as a gravitationally bound open cluster for many millions or even billions of years.

Such efficient conversion of interstellar gas into stars requires extraordinary environments which are not commonly found in most molecular clouds or star-forming complexes. The parent cloud must have a very large mass and density to survive the disruptive effect of stellar outflows or the birth of

Figure 8.4. One of the Pleiades stars, Merope, is associated with a small nebula seen in this image from HST. Radiation pressure from the star separates dust particles according to size and develops the streamers from the nebula. (G.H. Herbig & T. Simon, NASA/STScI).

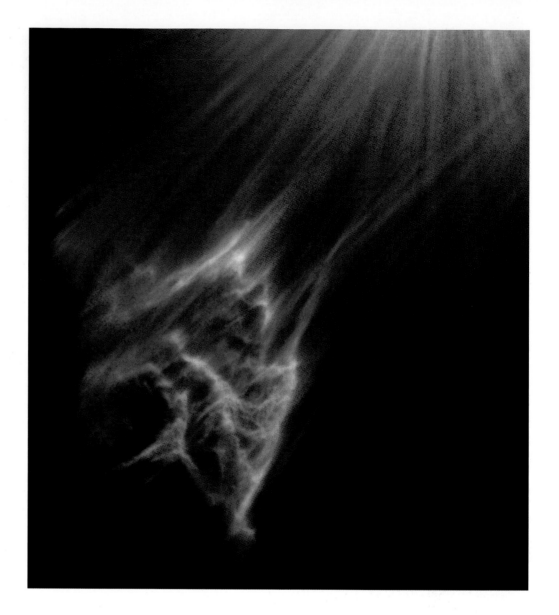

high-mass stars. Conditions like that exist only in the densest cores, which constitute just a small fraction of the volume of a giant molecular cloud.

As we will see in Chapter 9, the birth of massive stars, those with masses in excess of eight solar masses, produce ionizing radiation which excites *HII regions*. Such regions are hot (about 6000–10 000 degrees) and their constituent hydrogen ions move about with velocities of about 10 km/s. When the escape speed from a cloud is low, the HII regions expand, creating bubbles of plasma which can destroy most of the parent molecular cloud. Furthermore, these massive stars explode as supernovae[3] within about 30 to 40 million years of birth. Thus, they are far more damaging to their environments than lower mass stars. The birth of even one such star can destroy a molecular cloud core and, in regions such as the Orion Nebula, the birth of a few massive stars may be sufficient to stop further star formation.

There are situations, however, where the parent cloud is sufficiently dense and massive to prevent the massive stars from stopping further star

Figure 8.5. The open cluster M11 contains several thousand stars that were born about 200 million years ago. (J.-C. Cuillandre/ CFHT).

Figure 8.6. NGC 4755 is an open cluster in the constellation Crux. It is often called the "Jewel Box" because of its bright multi-hued stars. (NOAO/AURA/NSF).

formation. If the cloud has so much mass that the gravitational escape speed is larger than the typical speed of ions in an HII region, then the destructive effects of massive stars can be "bottled up." To see why this is so, consider an O star in a cloud core that has a large gravitational escape speed. A typical star will move about with a comparable velocity. As the O star moves around in the cloud, it will continuously encounter fresh dense gas which consumes its UV radiation. In such very dense regions, the O star cannot damage much of the cloud; its influence will be confined to a compact HII region in its immediate vicinity that follows it around inside the cloud. The gas that is left behind by the wandering star is no longer subjected to its UV radiation and will promptly recombine. When the effect of O stars is restricted in this manner, stars may continue to form in the cloud despite the presence of massive stars until a sufficiently high star formation efficiency is achieved to produce a bound cluster. The cloud may be destroyed and star formation brought to a halt only when one or more of the O stars explode as a supernova.

Super star clusters and globular clusters

Even more spectacular than open clusters, are the *globular clusters*: spherical swarms of stars containing 100 000 to over a million members packed into regions less than 10 light years in diameter (Figure 8.7). Our Milky Way galaxy contains slightly more than 100 globular clusters and all are ancient. Most date back to the time when our Galaxy formed, about 10 billion years ago. Consequently, only long-lived and ancient stars less massive than the Sun remain in these majestic clusters. The brightest globular cluster, known as

Figure 8.7. The globular cluster M3 contains about 500 000 very old stars and is located at a distance of 30 000 light years. (NOAO/AURA/WIYN).

Omega Centauri, looks like a faint and slightly fuzzy "star" to the naked eye. Large telescopes resolve this object into a spectacular myriad of long-lived, low-mass stars.

The formation of a globular cluster must require truly exceptional conditions that do not exist anywhere in the Milky Way today. When the loss of massive stars because of natural stellar death and the "evaporation" (see next section) of low-mass stars is considered, it becomes apparent that globular clusters must have been considerably more massive in the past. The progenitor molecular cloud must have had enough mass to leave behind millions of stars.

The cloud must have been much denser and more massive than any clouds seen in the Galaxy today. Furthermore, most of the mass of such a cloud must have been involved in one, cathartic episode of star formation: a *starburst* producing millions of stars in a time span of no more than a few million years. The parent clouds which gave birth to today's globular clusters may have had sufficient mass and pressure not only to resist the destructive power of hundreds of massive O-type stars, but even to confine the first few supernova explosions that such a cloud must have suffered.

As will be discussed in Chapter 15, studies of nearby galaxies show that, in the absence of an actual galactic collision, the most virulent starbursts occur either in galactic nuclei or in the somewhat disorganized gas-rich "late-type" spirals or irregular galaxies.[4] For example, the star-forming region known as 30 Doradus in the Large Magellanic Cloud (our nearest extragalactic neighbor) produced a cluster of about 400 massive O stars during a recent starburst. The 30 Doradus region is surrounded by a 3000 light year diameter superbubble indicating that dozens of massive stars have already died in this star-forming complex. For comparison, the most intense starburst regions in the Milky Way have spawned about 100 massive stars. Thus, the 30 Doradus complex is a far more active site of star formation than any in our Galaxy (see further discussion in Chapter 9). A region similar to 30 Doradus is located in a nearby late-type spiral galaxy known as M33 (the Triangulum galaxy located about 2 million light years from us). This region, known as NGC 604, spawned hundreds of massive stars in the last few million years. Most likely, none of these regions will leave behind globular clusters since none have formed in nearby galaxies for nearly 5 *billion* years. At best, these current starbursts will produce rich open clusters. Enormous as they are, such nearby starbursts are dwarfed by the super star clusters of the irregular galaxy known as He 2-10 located more than 200 million light years from us. In this object, astronomers have found several young starburst regions that have each spawned more than 5000 massive stars. Perhaps these extreme but distant star forming regions are forming globular clusters today.

It is possible that the largest globular clusters in our Galaxy may be the left-over remnants of the cores of dwarf elliptical galaxies[5] that merged with the Milky Way billions of years ago. The largest globular cluster in the Milky Way, Omega Centauri, contains over a million stars. Unlike most other globular clusters, it is slightly flattened by rotation. As will be discussed further in Chapter 15, it is quite likely that in its youth, the Milky Way swallowed a number of dwarf galaxies that happened to wander too close. The scrambling of orbits during the merger would have stripped stars from the outer parts of the incoming dwarf galaxy. These stars would eventually join the spheroidal bulge and halo[6] of the Milky Way. If the infalling galaxy interacted predominantly with the disk of the Milky Way, the star-rich core of the dwarf elliptical galaxy may have survived intact. In this hypothesis, we simply push back the birth of extreme globular clusters such as Omega Centauri to the distant past when the nuclear region of a dwarf galaxy was assembled. This begs the question, how are the stars in galactic nuclei and bulges formed? This issue will be discussed in Chapter 15.

Observations show that high pressure supergiant molecular clouds may form in merging galaxies. Figure 8.8 shows an image obtained with the Hubble Space Telescope of two galaxies in the process of colliding. As the

Figure 8.8. The collision of two galaxies, known as the Antennae, has triggered huge bursts of star formation, leading to the formation of numerous large star clusters, as seen in this HST image. (NASA/STScI).

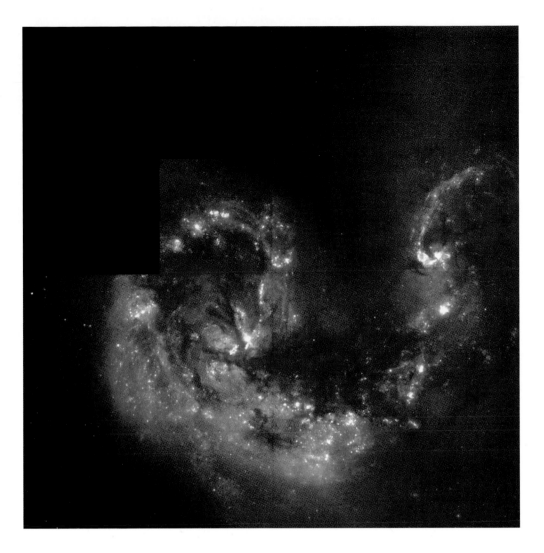

molecular clouds of the two galaxies plow into each other, violent star forming events have lit up in many places. Closer analysis of the resulting populations of young stars suggest that they may eventually become globular clusters. The UV production and radio emission of these starburst regions indicate that thousands of massive stars are forming. If the mass spectrum of forming stars is similar to that of more ordinary star forming regions, then the inferred number of low-mass stars forming in each starburst region must be in the millions. The super star clusters forming in nearby galaxy mergers may be modern-day analogs of the starbursts that may have spawned the Milky Way's 100 or so globular clusters.

The life and death of a cluster

Soon after birth, most stars diffuse away from their birthplaces and join the general stellar population of the galactic disk as isolated stars or binaries. The same fate awaits most of the stars that we now see in open clusters. Despite being bound by gravity, as clusters evolve, they too eventually lose

Figure 8.9. The planetary nebula NGC 6751 is an envelope of gas expelled several thousand years ago from the hot star visible at its center, which is nearing the end of its life. (NASA/STScI).

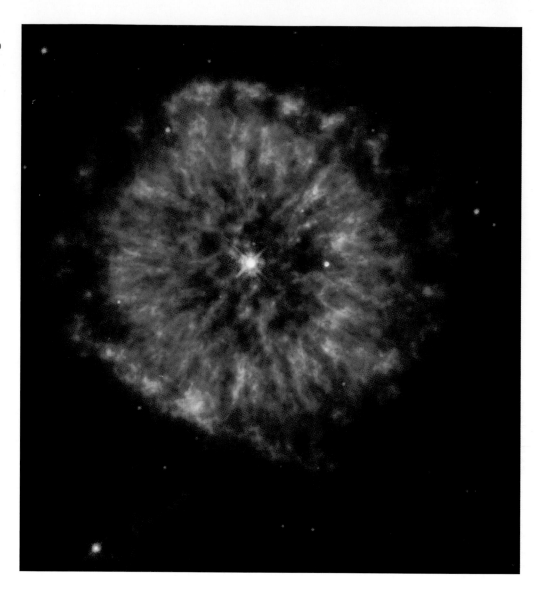

their stars. There are several processes that ultimately lead to the destruction of all open star clusters.

First, the most massive members of a cluster, those with masses over eight times the Sun, explode as supernovae. Their expanding gaseous remnants are expelled from the cluster. After about 40 million years, no stars massive enough to explode as supernovae remain. On average, the explosive demise of its most massive members removes almost 20 percent of a cluster's mass. As the cluster ages, stars with masses between a few and eight solar masses evolve into red giants and also expel most of their matter into space, though not nearly as violently as supernovae. Even stars like our Sun eventually run out of hydrogen fuel, and will eject their outer layers, revealing their stellar cores. The UV radiation emitted by these barren stellar cores can ionize the expanding outer shell of the star, producing the *planetary nebulae*.[7] Figure 8.9 shows the planetary nebula NGC 6751 as imaged with the Hubble Space Telescope. Over a billion years or so, loss of mass through planetary nebulae deprives open clusters of an additional 30 percent of the initial matter. Continued loss of mass through stellar death weakens the pull of the cluster's gravitational field. Weakly bound stars will tend to drift away as the cluster's mass declines with age. In addition to losing its most massive stars to old age, the cluster will also lose some of its least massive members as the ties that bind them weaken over time.

Second, stars in clusters wander around in a crowded environment, and from time to time have close encounters with other cluster members. This leads to an exchange of energy between stars, causing more massive stars to sink towards the cluster center and the lighter ones to migrate to the outskirts. Occasionally, as in the case of forming multiple star systems, three-way encounters eject the least massive members from the cluster. As a cluster slowly loses stars, it is said to be "evaporating."

Third, random encounters with giant molecular clouds and other star clusters exert tidal forces which accelerate the stripping of loosely bound low-mass stars from the outer parts of a cluster. Such encounters alone are sufficiently frequent that most open clusters in the Milky Way dissolve within a billion years. Weakly bound clusters can dissipate much faster. The removal of individual stars from clusters is one of the main mechanisms which populates the thick disk of stars in our Galaxy.

In combination, these processes tend to destroy the open clusters of our Galaxy within a few hundred million to a few billion years of their birth. However, the much more massive globular clusters can survive much longer since most have ages of the order of 5 to 13 billion years. Because they contain so many stars and are so tightly bound, globular clusters outlive open clusters. Only those globular clusters that repeatedly passed too close to the very dense clouds, clusters, and star fields found in the inner portions of our Galaxy (within about 10 000 light years of the nucleus) are thought to have been disrupted.

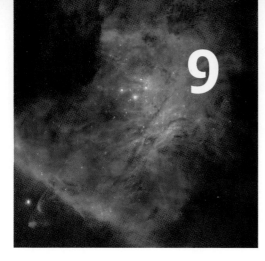

9 Chaos in the nest: the brief lives of massive stars

For most of their lives, stars evolve by converting hydrogen into helium in their cores. This thermonuclear fusion reaction replenishes the energy lost at the star's surface in the form of starlight. Consequently, the more luminous a star, the faster it consumes hydrogen. The most massive stars blaze with the light of a million Suns. Therefore, their fusion reactions consume hydrogen a million times faster than the Sun. Since these stars only contain a hundred times more hydrogen, these spendthrifts deplete their supply of fuel, living only a ten-thousandth of the lifetime of our Sun. While the Sun is expected to live for about ten *billion* years, the most massive stars survive only for about a *million* years.

Massive stars: live fast, die young

High-mass stars not only live fast, they also form fast. In contrast to lower mass kin such as the Sun, which take many millions of years to reach adulthood, massive stars can mature in less than a hundred thousand years. As discussed in earlier chapters, low-mass stars tend to shed their natal environs well before reaching the main-sequence stage, which represents the commencement of stellar adulthood. On the other hand, massive protostars mature with such rapidity that they usually reach adulthood still enshrouded within their placental clouds. In other words, they start life in the curious situation that they behave like adult stars cloaked in the environment of a newborn star. It is therefore very difficult to study infant high-mass stars since they are usually obscured within their natal dense cloud cores.

There are additional differences between high- and low-mass protostars. By the time a growing protostar reaches a mass of about 20 Suns, its own brilliance can halt the further inflow of matter from its birthing envelope. Thus, *radiation pressure* can stop stellar growth. Yet, Nature produces stars with masses up to about 100 times the mass of the Sun.

How can such massive protostars continue to grow despite the outward force of radiation? One possibility is that growth is fueled by mass flowing from the cloud onto the outer rim of a large accretion disk. Here, the infalling gas can be shielded by the inner portion of the disk. This matter then eventually spirals onto the massive star.

Another possibility is that the gas is accreted in such a dense state that most of the infalling mass is shielded from the forming star's light. Consider dropping a feather and a rock from a bridge. Compared to the rock, the feather will descend slowly. If an upward current of air catches it, the feather will rise against gravity's pull and be blown away. But, ignoring the breeze, the rock continues its plunge. Similarly, compact blobs of matter can accrete onto a luminous massive star while more tenuous clouds will be blown away by the intense starlight.

Ultimately, as the star's brilliance grows, the infalling envelope is halted and the gas is driven away by the pressure of light. As a result, an evacuated cavity forms in the envelope of the maturing massive star. The stream of expanding gas will ram into the infalling material further out, producing a dense shell at the outer boundary of the cavity. As infalling matter accumulates, the shell becomes completely opaque to the star's light. The opposing forces of light and gravity then fragment the shell into dense and opaque blobs surrounded by relatively transparent voids through which starlight can escape into the infalling envelope. The tug-of-war between the outward push of light and the inward pull of gravity now shifts in gravity's favor. Dense knots descend through the cavity and settle onto the star or its disk.

However, some researchers have a very different view of how massive stars are born. They suggest that the highest mass stars grow by swallowing lower mass sibling stars and their accretion disks. Young massive stars are usually found in the midst of dense crowds of low-mass stars. Furthermore, as discussed in previous chapters, protostars are surrounded by large accretion disks. And the younger the star, the more massive the disk. The youngest protostars often have disks which contain nearly as much mass as the protostar itself. If two such protostars happen to pass close to each other, their disks may get entangled, leading to a collision between the stars themselves. Crowding and the presence of massive disks makes it possible for the highest mass stars to grow by *cannibalizing* smaller protostars and disks that ramble too close.

The possibility of protostellar collisions may seem surprising since the vast distances between the stars in the sky imply that close encounters are extremely rare; so rare that collisions between ordinary stars in the Milky Way may never occur in the entire history of our Galaxy. But conditions in the dense star forming regions where massive stars are born are entirely different. In such regions, the spacing between young stars is often less than *one-hundredth* of the spacing of stars near the Sun.[1] Furthermore, the accretion disks of protostars are thousands of times bigger than the stars. These factors imply that in massive star-forming regions, the typical separations between protostars are frequently only slightly larger than the sizes of their accretion disks. Under these circumstances, protostellar collisions become not only possible, but likely.

When a passing sibling star wanders close enough to a massive protostar for their disks to be distorted by gravity, the encounter can lead to a merger. First, intense tides disrupt the disks, flinging some material out of the system. Then the speed of the incoming star is drained away by this circumstellar debris. Initially, the star may be captured into an orbit about the more massive object. But eventually friction may cause the massive protostar to absorb the mass of the hapless wanderer. Any remaining circumstellar debris eventually settles into a new disk around the massive object. This process may occur again and again until all protostars that wander too close to the massive objects are swallowed. The accretion of a solar mass object onto a 20 solar mass protostar can release about 0.1 percent of the energy released by a typical supernova explosion. It takes our Sun 100 million years to produce this much energy.

Hot bubbles, silverlined clouds and elephant trunks

Massive stars are hot. Consequently they emit copious amounts of ultraviolet radiation. As soon as massive stars form, they therefore begin to drastically alter their environment. Their ultraviolet light destroys the surrounding molecular cloud, creating in its place a hot bubble of ionized gas, and *ionized nebula* (Chapter 8). On color photographs such nebulae are visible as translucent glowing clouds dominated by red hydrogen emission.

The bubbles of hot plasma produced by massive stars expand into their surroundings. But the growth of a nebula depends on the structure of the cloud into which it moves. And molecular clouds contain dense clumps, filaments, much lower density cavities, and odd shaped voids. The resulting nebular expansion will be fast towards regions of low density, and slow where it encounters high density gas. As a result, nebulae develop remarkably intricate and often beautiful shapes.

Expanding ionized nebulae drive shock waves into surrounding cold gas. As the shock races ahead of the ionization front with speeds of a few to 10 km/s, gas is swept up, compressed, and accelerated. When molecules are pushed closer to the ionization front, the growing intensity of UV light dissociates and heats the medium. Eventually, these atoms flow through the ionization front, are ionized, and become incorporated into the growing HII region. Thus expanding ionized nebulae are usually preceded by dense shells of gas swept up from the surrounding molecular cloud. Over the course of millions of years, the shells can sweep through and destroy an entire giant molecular cloud; all because of the intense UV radiation of a few massive stars.

Occasionally a growing nebula runs into a dense condensation of gas which nearly stops its expansion in a particular direction. The dust and gas in such clumps block the ultraviolet light emitted by the nebula's high-mass stars, and matter in the shadow of the clump will remain relatively unaffected, at least for a while. As the expanding nebula races past into the lower density gas surrounding such clumps, a pillar of dusty molecular cloud material may survive in the shadowed region. The resulting protrusions point toward the hot stars in the nebula's interior. Astronomers have dubbed these structures *elephant trunks*. Figure 9.1 shows the famous "Pillars" in the Eagle Nebula. Ultraviolet light is eating into the leading surfaces of several clumps, producing glowing clouds. These silverlined rims signal the flow of freshly created plasma from the dense clumps at the tops of the pillars. Gradually these elephant trunks lose matter to the surrounding hot bubble. Figure 9.2 shows an infrared image of the same region. Light at infrared wavelengths can penetrate all but the densest cloud cores, allowing background stars to shine through.

Eventually, even the shaded portions of such protrusions erode. This is because the interior of an ionized nebula scatters the light of its exciting stars, like a steam-filled bathroom diffuses the light from a lamp. Thus, matter in the shadows of dense cores and opaque disks becomes exposed to scattered light and is ionized. Eventually, the dense clumps become detached and isolated from the rest of the cloud. Trapped in the interior of a nebula, they are doomed to evaporate unless the illuminating massive stars die first. Such clouds become isolated globules rimmed by glowing ionization fronts. Sometimes the globules lie directly in front of a nebula so that, from our vantage point, we see the un-illuminated back side of the clump in silhouette against its own glowing front (Figure 9.3).

Within a few million years of their birth, the massive stars responsible for ionizing a nebula will start to die. As they do, the bright rims of surviving globules and pillars fade, but the condensations sometimes remain visible in silhouette against background star fields or more distant nebulae. When a globule completely outlives its host OB association, it may survive for a very long time as an isolated dark cloud. This is most likely to happen if the cloud has managed to retain enough mass to be bound by its own gravity. Such isolated small dark clouds often contain no more mass than the Sun. The *Bok globules* discussed in Chapter 3 may have formed this way.

An overview of the Orion region

The constellation Orion contains the nearest region of massive star formation, and we have referred to it in passing in the previous chapters. Here we will examine this region and its history in more detail. Young massive stars dominate the naked eye appearance of this constellation, including the brilliant stars Betelgeuse in the upper-left corner of Orion and Rigel in the lower-right (Figure 9.4). Betelgeuse is a cool, red, and very luminous supergiant star whose enormous size would swallow the Earth's orbit if it were to replace the Sun. In contrast, Rigel is a blue supergiant which glows with dazzling brilliance. The three stars that form the distinctive Orion's Belt are also

Figure 9.1. The large elephant trunk structures in the HII region M16 are sculpted by the UV radiation from luminous massive stars. (NASA/STScI).

Figure 9.2. The elephant trunks in M16 become partly transparent when viewed at infrared wavelengths. (M. McCaughrean/ESO).

Figure 9.2. The elephant trunks in M16 become partly transparent when viewed at infrared wavelengths. (M. McCaughrean/ESO).

supergiant stars. Figure 9.5 shows an optical wide-field image of Orion and the Milky Way passing next to it. Figure 9.6 shows approximately the same area but seen in the mid- and far-infrared with the IRAS satellite. Whereas the optical image shows the young stars that already emerged from their clouds and the faint reddish glow of ionized hydrogen, the infrared image probes the embedded star forming regions and the regions of warm dust heated by newborn stars.

The most spectacular ionized nebula in the sky, the Orion Nebula, is located just below Orion's Belt stars.[2] The Nebula is a dense bubble of hot hydrogen plasma carved from the near-side of the Orion molecular cloud. The visual appearance of the Orion Nebula through a telescope is dominated by a gossamer tapestry of sheets, filaments, and clumps of blue-green light. Blazing like diamonds, seven brilliant blue stars light up the Nebula: three in a line below Orion's "Bright Bar" and four in a spectacular quadruple system, the *Trapezium*, in the center complete the celestial masterpiece (see Figure 9.7). The Trapezium stars may be less than a hundred thousand years old, and star formation is not yet complete in this region. Just behind the Orion Nebula, several very luminous infrared sources are embedded in

Figure 9.3. Thackeray's Globules in the HII region IC 2944 exemplify the destructive effect of luminous massive stars (in the background), which tear the cloud apart. (B. Reipurth, NASA/STScI).

a massive molecular cloud core. A large explosion occured in this region about a thousand years ago. Although no astronomers recorded this event, we see today an expanding field of interstellar shrapnel flying out at speeds of hundreds of kilometers per second. The pieces leave behind glowing trails of heated molecular hydrogen resembling a fireworks display (Figure 9.7). A high-mass protostar with a luminosity of nearly 100 000 Suns lies at the center of the expanding debris. In the explosion, the star released 0.1 percent of the energy produced by a supernova. Some astronomers think this eruption was generated by a scaled-up version of the outflows seen in lower mass stars and discussed in Chapter 6. However, it is also possible that this event was produced by an act of protostellar cannibalism, as described earlier.

The Orion Nebula merely marks the brightest and most recent active region of massive star formation in Orion. The nebula sits like a jewel near the northern tip of a giant molecular cloud which covers most of the southern part of the constellation. Known as the Orion A cloud, it is still producing stars. The Orion A cloud resembles a giant comet nearly 40 light years long. Its northern end is dense and compressed, while its large south-eastern

Figure 9.4. The constellation Orion represents a hunter. Underneath the three belt stars hangs a sword, which contains the Orion Nebula. (Wei-Hao Wang).

part widens with distance and is more diffuse than the northern part. The south-eastern part is just now starting to form its first stars.

North of the Orion Nebula lies a chain of collapsing cloud cores that contain the greatest concentration of protostars discovered anywhere in the sky. A bit further north, above the northern tip of the Orion A cloud but

Figure 9.5. A wide-angle view of the entire constellation of Orion and its surroundings. (Wei-Hao Wang).

still within the sword, lies an older and less well known ionized nebula born from the Orion A cloud.[3] Thousands of stars have formed from the dense northern part of Orion A.

Orion also contains other star forming clouds. A second giant molecular cloud known as the Orion B cloud lies north of the Orion A cloud and east of Orion's Belt. An ionized nebula and several young embedded star clusters mark centers of ongoing star formation here.[4] The western rim of this cloud is lit up by the intense light of Sigma Orionis, a massive star at the center of an expanding cluster of hundreds of stars formed about two million years ago. Already, the radiation from this star has completely destroyed the original cloud from which this cluster formed. All natal envelopes and most circumstellar disks have been destroyed. Therefore, the members of this star cluster are all visible. The current edge of the molecular cloud is

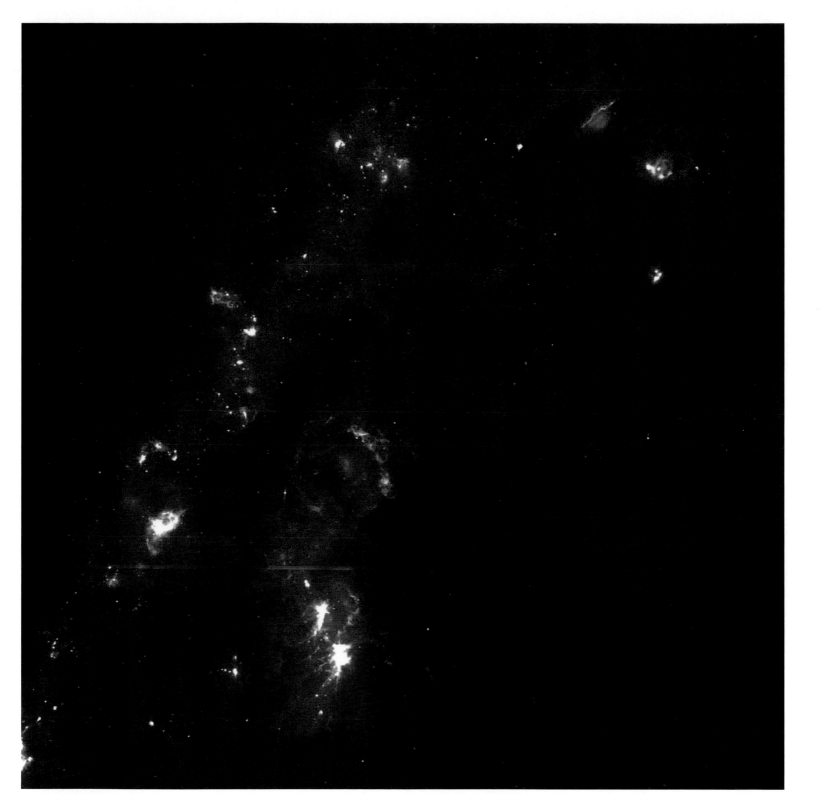

Figure 9.6. The approximate same area as Figure
9.5 seen at mid- and far-infrared wavelengths. The
image reveals large complexes of cool dust.
(Th. Preibisch).

Figure 9.7. The fingers of shock-excited molecular hydrogen emission (red/orange) indicate that an eruption occurred about a thousand years ago in the vicinity of one of the massive stars still forming from the molecular gas located immediately behind the Orion Nebula. (Subaru/HST).

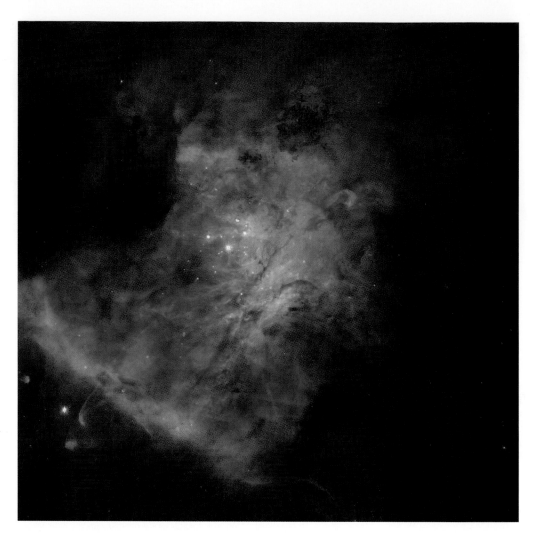

by now located more than 10 light years east of Sigma Orionis. Here, the corrugated cloud edge of Orion B is the site of other cloud cores which are actively spawning stars. One of these is the famous Horsehead Nebula (see Chapter 4).

In order to understand the complicated history of star formation in Orion, we will now examine in more detail the chain of events that are started when a young massive star is born.

When star death triggers star birth

Clouds trapped in the interiors of ionized nebulae are sometimes crushed by the pressure of the surrounding plasma. As starlight illuminates a cloud, ionization of its skin generates powerful inward pressure which compresses the cloud. Eventually, as the interior density grows, the force of gravity may overcome the cloud's resistance to compression, resulting in gravitational collapse. In contrast to quiescent star formation where cloud cores collapse on their own (Chapter 4), illumination by a nearby high-mass star can drive an implosion leading to rapid star birth. This process is called *triggered star formation*.

In many OB associations, the birth of high-mass stars and the creation of hot plasma bubbles can trigger the formation of additional massive stars in dense gas just inside the nebular rim. As the dense, shocked gas that advances into a cloud ahead of the ionization front is accumulating ever more mass, it can become unstable to gravitational collapse. Thus expanding HII regions can induce star formation in their shells. Alternatively, the advancing nebula can overrun pre-existing cloud cores and push them beyond the threshold of collapse. This process can occur when the compression produced by the advancing nebula triggers the wholesale collapse of a relatively massive portion of a surrounding cloud. Perhaps the high-mass stars now forming behind the Orion Nebula were triggered in this fashion.

Eventually, as these second generation massive stars emerge from their birthplaces, they will create their own bubbles. And further star formation may be induced in the cloud. Like a wildfire passing from tree to tree, triggered massive star formation may pass through the surrounding molecular cloud in a process known as *propagating* or *sequential* star formation.

The compression produced by an expanding nebula is not the only force capable of triggering star formation. As the most massive stars in an OB association evolve and die in supernova explosions, these blasts can also dramatically compress clouds in their path. However, if a cloud too close to the explosion is hit by such a blast, it may be shattered. When the remnants of an exploding star encounter interstellar matter, the speed of the outrushing material gradually diminishes. As the shell of swept-up matter slows, the impact on more distant clouds can be considerably more mild. Rather than dispersing the cloud, the resulting shock may be gentle enough to just compress it. Under the right conditions, gravitational collapse may again set in, resulting in a new generation of stars induced by the death of a massive star.

A history of star birth in Orion

The Orion constellation harbors two giant star forming clouds, which have given birth to many *tens* of thousands of stars. These members of the Orion OB association provide a fossil record of star formation over the past 15 million years when star formation first ignited there.

Over the years, Orion has produced a whole sequence of expanding clusters. Each is known as a *sub-group* of the Orion OB association. These groups trace episodes of active star formation interspersed between periods of relative quiescence. The locations and ages of these clusters trace the history of propagating star formation in Orion.

Star formation in Orion first ignited north-west of Orion's Belt 15 million years ago. Long-dead massive stars swept this region clean of interstellar gas, leaving behind an expanding cluster of medium- and low-mass stars. But as this so-called Orion 1a subgroup compressed the surrounding cloud, star birth was again triggered a few million years later to the south-east. This new burst of star formation gave birth to Orion's Belt and many thousands of low and intermediate stars that comprise the Orion 1b sub-group.

At this point, propagating star formation bifurcated at Orion's Belt. One branch of star formation propagated south towards the Orion A cloud. Roughly five million years ago, stars started forming about five degrees south of the Belt, including the faint naked eye stars that today make up most of Orion's Sword.[5] These stars are known as the Orion 1c sub-group and lie directly in front of the Orion Nebula.

The Orion 1c sub-group compressed the northern end of the Orion A cloud lying in the background. It triggered a burst of star formation that gave birth to the Orion Nebula and the current crop of highly embedded protostars in the dense northern filament in Orion A. Within the next few million years, the more diffuse southern portion of Orion A may be compressed as the Orion Nebula itself evolves and triggers star formation in the remnants of the Orion molecular cloud. Thus, the Orion Nebula itself may ignite future bursts of star formation. As the Orion Nebula fades, new sites of massive star formation may occur in the dark south-eastern reaches of today's Orion A cloud, perhaps forming the final sub-group of the Orion OB association as the remaining gas in Orion A is dispersed.

The second branch of propagating star formation progressing from the Orion 1b sub-group (the Belt Stars) moved east towards Orion B. About two to four million year ago, this branch produced the stars of the Sigma Orionis sub-group. Today, star formation here is propagating further east and north-east. Several very young clusters of stars are still forming within the remaining portions of Orion B. But, like in Orion A, star formation in Orion B will also soon disrupt the remaining gas. Here, too, star formation is predicted to come to an end within the next few million years.

Orion blows a bubble

Along with large numbers of garden variety low-mass stars, each of Orion's sub-groups nurtured dozens of high-mass ones. Some are now evolving towards stellar death. They are the giant and supergiant stars which comprise Orion's brightest naked eye stars. But many even more massive stars spawned by the sub-groups have already died. Their radiation, stellar winds, and supernova explosions have collectively inflated an enormous *superbubble* 300 by 900 light years in diameter. The vast low density interior of the superbubble is filled with an ultra-hot plasma glowing faintly at soft X-ray wavelengths. Known as "Orion's Cloak" this bubble extends from the eastern edge of Orion, all the way to the constellation of Eridanus in the west (Figure 9.8).

As it expanded, Orion's superbubble overran the surviving parts of its original parent molecular cloud. The resulting compression helped to trigger

Figure 9.8. A superbubble known as "Orion's Cloak" extends around the many OB stars in Orion. (M. Bissell & R.S. Sutherland).

new bursts of star formation. In turn, the massive members of these new stellar groups contributed their own energy to the growth of Orion's Cloak. In fits and starts, star formation propagated from the north-west of the Belt Stars towards the south and east. The Orion A and B clouds are the surviving relics of Orion's original giant molecular cloud.

In addition to the large Orion A and B molecular clouds, many smaller clouds inhabit the interior of Orion's Cloak. Light from Orion's massive stars, and the outflowing plasma from the superbubble's interior, have sculpted them into comet-shaped clouds, with their dense heads and diffuse tails streaming away from the center of Orion (Figure 9.9).

As it plows into the surrounding interstellar space, the expanding Orion superbubble has swept up immense quantities of gas and dust. The glow of this warm dust is visible in the far-infrared wavelength portion of the spectrum. In recent decades, several NASA satellites have obtained pictures of Orion's glowing shell from space, and pictures obtained with radio telescopes on the ground reveal that this giant shell is expanding. The shell rims facing

Figure 9.9. A cometary-shaped cloud points towards the massive stars in Orion. Their influence most likely triggered the birth of a young star driving a large HH flow emanating from the globule head. (J. Bally & J. Walawender).

Orion's OB association are ionized by surviving massive stars, and can be seen as a giant filamentary nebula of glowing hydrogen plasma.

The ejected runaway stars

A fraction of the massive stars formed in OB associations move with velocities high enough to escape from the cluster, sometimes with speeds of hundreds of kilometers per second. The stars AE Aurigae and Mu Columbae are two examples of *high velocity runaway stars*. They are moving directly away from Orion with speeds of 137 and 141 km/s, respectively. When their motions are traced backwards, one sees that the two stars were ejected from near the Orion Nebula about 2.6 million years ago. They are now located many tens of degrees and hundreds of light years from their point of origin. About 10 percent of the massive stars in an OB association are ejected with such high velocities, and another 20 or 30 percent move away with lower speeds of the order of 10 km/s. The star Betelgeuse in Orion is an example of such a low-speed runaway. This red supergiant in the north-east corner of Orion is moving towards the north-east. It apparently originated somewhere between us and the Orion 1a subgroup just north-west of the Belt Stars in Orion about 10 million years ago.

How can some stars be launched from their point of formation with such high speeds? Two mechanisms have been proposed: the breakup of close binaries when the most massive member explodes as a supernova, and the dynamical ejection from dense clusters. Observations show that binaries and multiple stars are more common among massive stars than among low-mass stars. Stars more massive than eight times the Sun die in supernova explosions about 30 to 40 million years or less after birth. The more massive the star, the sooner it explodes. In a close binary consisting of a pair of massive members, the more massive star will explode first. The demise of the massive star will unbind the orbital motion of the surviving companion which, like in a sling-shot, flies away with its former orbital velocity.

Dynamic interactions similar to those that rearrange non-hierarchical multiples into hierarchical ones (Chapter 5) occasionally eject a massive star. Recent data suggest that this is how the above mentioned massive stars AE Aurigae and Mu Columbae were ejected from the Orion region. Apparently the ejection involved four stars in the 1c sub-group. As a result of their inter-action, one tight binary was formed and two stars were ejected as runaways. The binary star has been identified as Iota Orionis, and its motion indicates that it was also located at the point of intersection of the two runaway stars at the same time they were there.

Runaway stars provide an important mechanism by which star formation can seed further star formation at large distances. As discussed above and in Chapter 14, supernova explosions may play a vital role in triggering the collapse of clouds. When a massive runaway star eventually explodes, it may trigger additional star formation in clouds located hundreds of light years from its point of origin.

Giant star-forming regions

As spectacular as Orion is from our nearby vantage point, it is a relatively minor star forming region. Far more active sites of massive star formation dot the Milky Way galaxy and its nearest galactic neighbors. For example, Figure 9.10 show images of the galactic star cluster known as NGC 3603. Though over ten times farther from us than the Orion Nebula, it is one of the more spectacular regions of star formation in the southern sky. Silver-lined elephant trunks point like arrows to a cluster of massive young stars centered in a glowing ionized nebula. This grouping of hot and massive stars is so rich and dense that, at first glance, one might think that it is a globular cluster. It contains over fifty high-mass stars and thousands of lower mass ones too dim to be seen individually. One member of the cluster, known as Sher 25, is about to die. This bloated blue supergiant has shed a ring of gas and a collimated outflow at right angles to the ring. Some time in the

Figure 9.10. The young galactic cluster NGC 3603 where over 50 massive stars were born several million years ago. (ESO/VLT).

future, it will light up the southern sky when it explodes as a supernova. The Rosette Nebula is another spectacular bubble structure created by a cluster of young massive OB stars (Figure 9.11).

The 30 Doradus region in the Large Magellanic Cloud, a small satellite galaxy orbiting our Milky Way, is an even more dramatic site of high mass star formation (Figure 9.12). This complex contains the largest and most massive star forming region in our entire local group of galaxies. Its filamentary nebula is nearly a thousand light years in diameter. Therefore, if it were located at the distance of Orion, it would fill nearly half of the sky. Its nebula would light up our nights like the full Moon. Its central star cluster contains hundreds of massive stars within a region only several light years in diameter, in addition to hundreds of thousands of newborn stars of much lower mass.

Figure 9.11. The Rosette Nebula is a large bubble excited by the UV light from a group of young massive stars. (T.A. Rector, NOAO/AURA/NSF).

Figure 9.12. The 30 Doradus star-forming complex in our nearest galactic neighbor, the Large Magellanic Cloud, as observed at visual wavelengths. (Nathan Smith).

Part II Planetary systems

10 Solar systems in the making

Knowing our own backyard: the Solar System

The Sun is a truly unremarkable star in all respects, except that it happens to be *our* Sun, the source of all of our light and energy, without which we would not exist. This lack of being special is important, because it suggests that whatever processes were involved in forming the Sun are likely to be universal and might apply to most stars. That a retinue of planets formed shortly after the birth of our Sun suggests that planet formation could be a natural by-product of star formation, and that planetary systems may be common. The flurry of recent discoveries of planets around many nearby stars lends credence to this view.

So what are the general properties of our Solar System, which any theory of its formation should be able to explain? It is likely that the Solar System evolved with time, so we cannot be certain that all of its present-day characteristics reflect conditions that reigned in the distant past when it formed. However, by considering only the most general characteristics of our Solar System, we can identify fundamental properties that reflect its origin.

First, our planetary system consists of three types of planets. Small, dense, and rocky objects, known as the terrestrial planets for their resemblance to the Earth, occupy the inner Solar System. Giant planets with much lower densities, also known as the gas giants, are found further out. Jupiter (Figure 10.1) and Saturn are gas giants. Finally, ice giants (Uranus and Neptune) are most distant from the Sun.

Second, all the planets move around the Sun in the same direction and in a more or less common plane, which is also approximately the equatorial plane of the Sun.

Third, planets rotate in the same direction as their orbits, but their spin axes have a large range of inclinations to their orbital planes.

Fourth, the Sun contains 99.9 percent of the mass but the planets contain 98 percent of the angular momentum of the Solar System.

Fifth, all giant planets have numerous moons, the larger of which orbit in the equatorial plane of their planet.

Sixth, the outer Solar System beyond the orbit of Neptune is occupied by a swarm of small bodies composed of water ice, various frozen gases, dust grains, and larger rocky objects.

Figure 10.1. Jupiter is the largest planet in the Solar System, and here dwarfs Io, one of its Galilean moons. (NASA/Cassini).

Seventh, the chemical composition of Solar System objects vary with distance from the Sun.

Numerous theories have been put forward over the centuries attempting to explain the origin and general features of our Solar System. The first attempts were made in the eighteenth century by the German philosopher Immanuel Kant and (independently) by the French mathematician Pierre

Simon Laplace, both of whom envisaged an initial *solar nebula* as a flattened rotating disk of gas and dust, out of which the planets condensed.[1]

This early concept of a flattened, rotating solar nebula has survived and gained increasing support by the discovery that T Tauri stars are often surrounded by circumstellar disks. Active research into the dynamical evolution and chemical composition of the early disk/nebula has significantly increased our understanding of the basic physical processes that controlled the formation of our Solar System. However, many open questions remain and despite great progress in recent years, planet formation is still not well understood.

Ice, dust, rocks, and planetesimals

As the solar nebula formed from a collapsing core in a molecular cloud, it was heated by several processes. As matter fell onto the disk forming around the proto-Sun, gravitational potential energy was converted into heat. Subsequent viscous dissipation in the rotating disk and the decay of radioactive elements released additional energy. While these sources deposited energy into the deep interior of the disk, the nascent Sun eventually irradiated and warmed the surface layers of the disk. Calculations show that the inner disk within a few AU of the Sun reached temperatures in excess of 2000K. In this region, dust grains were vaporized so that the inner disk was completely gaseous. But temperatures were lower at larger distances from the proto-Sun. In the outer reaches of the solar nebula, tens to hundreds of AU from its center, interstellar molecules, grains, and even ices survived.

With time, the solar nebula gradually cooled and permitted the re-formation of molecules and solid particles. Each species of molecule and type of grain has a characteristic temperature at which it forms or condenses. Weakly bound molecules and ices can only form in the colder outer reaches of a disk. But more robust species such as the silicates and various iron-bearing compounds can form in the warmer inner disk. Condensation in the presence of a temperature gradient led to systematic variations in the chemical composition of the solar nebula with distance from the young Sun.

The gradual formation of solid particles in the cooling disk had a profound influence on the subsequent evolution of the solar nebula. In the inner nebula, the first important condensates were silicates and iron compounds. In the outer, cooler regions, vast quantities of carbon dioxide, water, and other types of ice accumulated and were mixed with smaller quantities of pre-existing interstellar grains inherited from the original cloud from which the Sun and its disk contracted. The makeup of the planets reflects these variations. While the terrestrial planets are mostly composed of rocky materials such as silicates and metals, the gas and ice giants from Jupiter and beyond contain vast quantities of hydrogen, helium, water, and other less refractory compounds.

Gentle encounters between grains eventually led to the formation of small, fluffy, and highly porous aggregates held together by electrostatic forces. In the cooler regions of the disk, atoms and molecules that collided with these particles stuck and contributed to their growth. As the condensates grew, they settled towards the disk mid-plane. Because grains of different sizes and densities fall with different speeds, collisions controlled particle evolution. While gentle encounters promoted grain growth, more violent collisions shattered fragile particles. But, on average, the competition between grain growth and destruction favored growth. Calculations show that under conditions likely to have existed in the inner solar nebula, the formation of centimeter-sized particles was very rapid, taking only about 10 000 years. Eventually, pebble-sized grains accumulated in the mid-plane of the disk and the abundance of solids became comparable to the gas, orders of magnitude higher than in a typical interstellar cloud.

As particles in the mid-plane grew to centimeter and meter scales, they started to drift in towards the forming Sun, and faced the potentially catastrophic fate of falling in. To understand this surprising result, we have to consider how large and small particles interact with gas in the disk. Gas in a protoplanetary disk orbits the central star slightly slower than the Kepler speed. This is because pressure gradients contribute to the balance between gravity and orbital motion; but the motion of particles less than a centimeter in diameter is strongly coupled to the surrounding gas. Like snowflakes blowing with the wind, collisions with atoms and molecules ensure that small bodies move *with* the gas. On the other hand, like large hail stones falling through the strong winds of a thunderstorm, particles larger than a few centimeters have so much mass and inertia that they try to move independently of the gas. Bodies larger than about a kilometer respond *only* to the force of gravity, and strictly obey Kepler's laws of motion about the Sun with virtually no effects caused by the pressure gradients.

As particles grew and decoupled from the gas and started to approach the Kepler speed appropriate for their distance from the Sun, they encountered a headwind of more slowly orbiting pressure supported gas and dust. This headwind drained orbital angular momentum and initiated an inward drift towards the young Sun. This effect peaked for centimeter- to meter-scale boulders which drifted towards the Sun with speeds approaching a million kilometers per year. At this rate, boulders can be dragged from a 1 AU orbit into the Sun in about a hundred years! Consequently, rocks with sizes ranging from centimeters to meters were at risk of being lost from the solar nebula.

What was the fate of these in-spiraling solids? Some probably did fall into the Sun while others accumulated at the inner disk edge. But much of this material survived to form the terrestrial planets. How? We do not yet have a clear answer. One possibility is that growth to the kilometer-scale was so rapid that it occurred on a time-scale even faster than infall. The kilometer-sized and larger bodies that somehow survived are known as *planetesimals*.

The birth of planets

Planetesimals, the building blocks of planets, had sizes ranging from 1 to over 100 kilometers. As these bodies swirled around the young Sun, the two most important processes for their evolution were gravitational interactions and physical collisions. The mutual gravitational tugs of distant encounters disturbed the orderly circular orbits of the planetesimals about the Sun and caused them to move on ever more elliptical paths. But, while planetesimals on perfectly round orbits never collide, ones on elliptical orbits do. The more elliptical the orbits, the more violent the collisions. If the relative velocities of colliding planetesimals are small, they may coalesce. But if the relative velocities become too large, the result can be catastrophic, with one or both bodies breaking up. Once more there was competition between growth and destruction.

Larger bodies have stronger gravitational fields and are therefore more efficient at sweeping up debris in their paths. As they grow, they become increasingly resistant to shattering and can better withstand large impacts. Computer models show that a sequence of ever more violent collisions occurred as growing planetesimals smacked into their siblings with ever greater force. Out of the millions of smaller bodies, collisional growth led to the emergence of hundreds of large planetesimals that eventually reached about the size of our Moon. But a system consisting of so many gravitationally interacting bodies is not stable. Within about 10 to 100 million years, collisions among the planetesimals lead to their coalescence into a handful of planetary embryos about the size of the terrestrial planets (Figure 10.2). The first few hundred million years was an extraordinarily violent epoch in the history of our Solar System.

The inner planets, Mercury, Venus, Earth, and Mars, are naturally accounted for in the processes described above. But how did the giant planets of the outer Solar System gain their titanic masses? The giant planets consist of mainly hydrogen and helium and contain large amounts of ice. Their compositions are much closer to that of the Sun than to the rocky planetesimals of the inner Solar System which contain very little hydrogen and helium. While the terrestrial planets grew by the coalescence of rocky materials, the giant planets must have scooped up additional material from the early solar nebula without discriminating between light gas and solids. There are two competing views among researchers on how this occurred.

In the "standard model" favored by most researchers, the cores of the giant planets formed by the collisional growth of planetesimals just like the inner planets. But once these cores grew sufficiently large, they also began to accrete surrounding gas rich in hydrogen, helium, and ices. In the inner solar nebula, the light gases hydrogen and helium may have been too hot to be accreted onto the forming rocky planetary embryos. Their gravitational pulls were not strong enough to accrete more than a token amount. Furthermore, by the time the embryos grew sufficiently large to accrete gas, the gas was

Figure 10.2. Collisions in the early Solar System were key to the formation of planetesimals and planets. (Painting by Dan Durda).

gone. It was either accreted by the Sun or ejected from the inner solar nebula by the young Sun's powerful stellar wind. But in the colder, outer solar nebula, the light gases were abundant. Furthermore, the large abundance of ices enhanced the growth rate of planetary embryos. They apparently grew massive enough to accrete large amounts of the outer nebula's abundant hydrogen and helium. As the mass of accreted gas exceeded the mass of the rocky cores, the increasing strength of the planetary gravitational field resulted in a process known as *runaway accretion*. Growth did not stop until all available gas in the vicinity of the giant planet's orbit was swept up (Figure 10.3). In the outer Solar System, the lower Keplerian orbit speeds imply a smaller

Figure 10.3. Computer graphics of a giant planet in the process of forming by sweeping up gas and forming a gap in the surrounding disk. (W. Kley).

amount of shear between adjacent orbits. Thus, planetary embryos of a given mass accreted more mass from a larger region than in the inner solar nebula.

This model explains the most important properties of the giant planets: their composition. Jupiter, Saturn, and Uranus/Neptune are enhanced in heavy elements relative to the Sun by factors of about 5, 15, and 300, respectively. The formation of ice/rock cores followed by runaway accretion of light gases can readily explain this pattern if the amount of available gas diminished with increasing distance from the young Sun.

Critics of the standard model argue that it may take too long to assemble the planetary embryos by collisional build-up to allow the subsequent growth of giant planets by a phase of runaway accretion. In many star forming environments with massive stars such as in the Orion Nebula, gas in the outer parts of circumstellar disks may be dispersed by a variety of processes in less than a million years, long before planetary embryos can grow enough to accrete gas from their environment.

An alternative model for the formation of giant planets has therefore been proposed. In this so-called "gravitational instability" model, giant planets form directly from the disk in a few thousand years. When a protoplanetary disk is young and very massive, its mass density may be so high that self-gravity of AU-scale regions exceeds the shear of Keplerian motion. Such a region can collapse in response to its own gravitational self-attraction, forming a bound object with about the mass of a giant planet. Figure 10.4 shows four different stages in a computer simulation of the formation of a gas giant from collapse within a very dense protoplanetary disk.

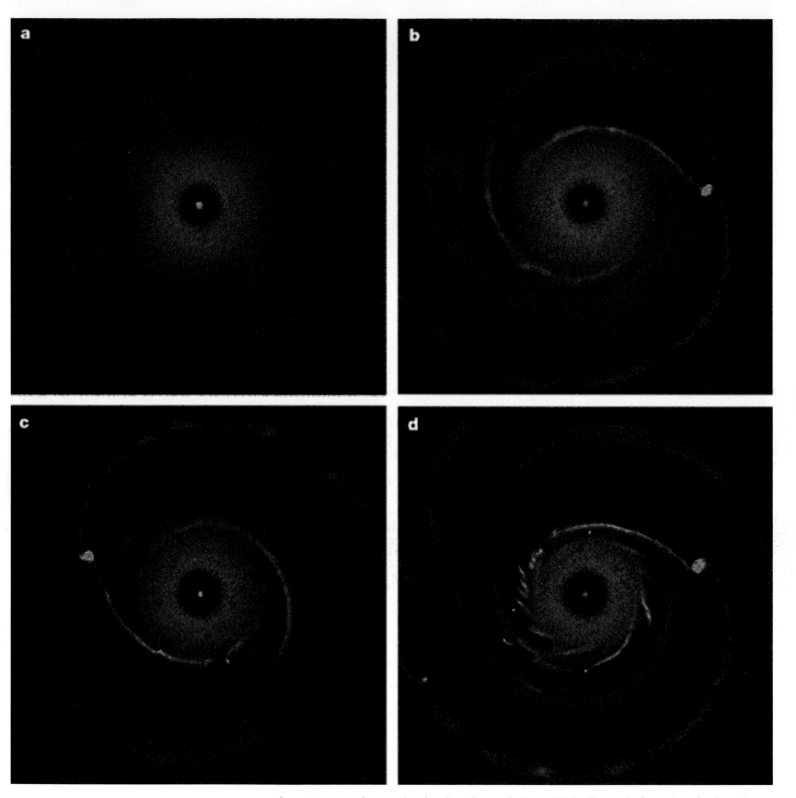

Figure 10.4. Four images showing the collapse of a dense disk leading to the formation of a giant planet. (P. Armitage & B. Hansen).

One difficulty with this model is that it does not easily explain the observed compositions of the giant planets in our Solar System. Gravitational instability is even-handed. All matter, regardless of composition, is dragged into the forming object. Ices, rock, hydrogen, helium, and every other component of a disk would be accreted. Thus, the resulting planets should reflect the original composition of the disk, which was close to that of the interstellar cloud from which it condensed, and similar to the composition of the Sun. Thus, the gravitational instability picture does not easily explain the observed abundances in Jupiter, Saturn, Uranus, and Neptune. To overcome this difficulty, proponents have to argue that when a disk goes gravitationally unstable, it has already been depleted of hydrogen and helium, and that this depletion becomes more severe with increasing distance from the proto-Sun. One possible depletion mechanism will be discussed in Chapter 12.

A second difficulty is that the formation of giant planets too early in the evolution of a disk may cause them to spiral into the proto-Sun. The sudden formation of a giant planet results in the excitation of a spiral wave in the disk that drains orbital angular momentum from the planet, causing it to migrate inward. The radial migration might stop if the giant planet reached a large gap in the disk, such as the central hole thought to exist around many protostars. Perhaps this is one way of forming the "hot Jupiters" to be discussed in Chapter 13.

In either scenario of giant planet formation, a young or still accreting giant planet is much larger than the present-day Jupiter. Models show that Jupiter must have been nearly an AU in diameter when it first formed. Shrinkage to its present size required the dissipation of heat and angular momentum. It may take nearly 100 million years for young giant planets to approach their mature dimensions. Even today, Jupiter emits some of its own internal energy. The predicted properties of very young giant planets have several interesting implications. First, as they radiated away their own gravitational potential energies, they emitted substantial amounts of infrared radiation. Second, their large dimensions enabled them to reflect a relatively large fraction of their parent star's light. For these two reasons, young giant planets could be considerably easier to detect over interstellar distances than mature planets. We will discuss the ongoing search for planets orbiting other stars in Chapter 13.

Moons and rings everywhere

In 1610, Galileo aimed his telescope at Jupiter and discovered that it was orbited by four moons. When he looked at Saturn, he found its majestic rings. He thus opened what would become the rich new field of Solar System research. In many ways, Jupiter's four Galilean satellites Io, Europa, Ganymede, and Callisto constitute a miniature planetary system with innermost Io being rocky and outermost Callisto being half rock and half water-ice.

They orbit near the equatorial plane of Jupiter and in the same direction as the spin of the giant planet. This suggests that the Galilean moons formed by accretion from a disk rotating in the equatorial plane of Jupiter shortly after its own formation.

It is now known that Jupiter has dozens of additional, but much smaller moons, many just a few kilometers across. These satellites tend to be in highly eccentric orbits, with most moving opposite to Jupiter's spin direction. These moonlets are probably leftover planetesimals that were captured by Jupiter's large early atmosphere[2] or possibly younger asteroids which were caught by three-way chance encounters with pre-existing moons. All of the Solar System's giant planets have similar groups of both regular satellites as well as captured irregular ones. Figure 10.5 shows an image of Phoebe, one of Saturn's small moons, as imaged by the Cassini probe orbiting Saturn. White streaks exposed along the walls of some craters suggest the presence of ample deposits of water ice. This is most likely a typical appearance of an icy planetesimal from the region of the giant planets.

The giant planets all have ring systems. The rings of Saturn are by far the largest and best known (Figure 10.6). They consist of particles ranging in size from dust grains to boulders to small moonlets. Even though Saturn's rings extend across several hundred thousand kilometers, they rotate in a single plane not more than tens of meters from top to bottom. This extreme flattening of the ring system occurs because the particles continuously collide, lose energy and re-distribute angular momentum. Planetary rings are sheared by strong tidal forces near their giant planets, so they may have been produced by the disruption of a moon that drifted too close to the planet. Indeed, ring systems are not likely to be primordial, but rather the result of a catastrophic event that may have occurred long after the birth of the planet.

The inner planets Mercury, Venus, Earth, and Mars only have three satellites altogether. Of these four planets, Earth has one moon and Mars two. Apparently, the small terrestrial planets were not surrounded by their own circumplanetary disks. The moons of Mars, called Phobos and Deimos, are tiny and could possibly be planetesimals that were captured by Mars. Our own satellite, the Moon, has a much more mysterious origin. It has a much larger mass relative to its parent planet than any other moon in the Solar System (with the exception of Pluto, see below). Many hypotheses have been proposed to explain its origin, but difficulties arise with its composition, which resembles the Earth's mantle except that it lacks volatile material. To explain that, consensus has shifted towards the protoplanet collision scenario.

About 4.45 billion years ago, when the Earth had nearly reached its present size, it suffered a violent collision with another planet or planetary embryo about the size of Mars. The impact destroyed the impactor and nearly broke the Earth apart. A plume of incandescent vapor and molten rock was ejected by the impact. Some of this material escaped, some fell back to

Figure 10.5. Saturn's small moon Phoebe is an ice-rich body overlain with a thin layer of rocky material. It is most likely very similar to planetesimals in the outer Solar System. (NASA/Cassini).

Earth, and the rest went into orbit. Most volatile materials were vaporized by the impact and lost to interplanetary space. Computer simulations show that such an impact can launch roughly one lunar mass of material into orbit around the Earth and that this debris rapidly coalesces into a single body. In the protoplanet collision scenario, the Moon was created from the mantles of the colliding planets whose volatiles (such as water) were lost to

Figure 10.6. The rings of Saturn are several
hundred thousand kilometers across, but have a
width of only about ten meters. (NASA/Cassini).

Figure 10.7. The Earth's Moon seen from an unfamiliar angle by the Clementine spacecraft. (NASA/Clementine).

space. Thus, the similarity of the Moon's composition to the Earth's crust, with the absence of volatiles, can be readily explained.

It might seem very unlikely for the young Earth to be struck by another planetary embryo. But violent collisions were commonplace in the young Solar System as numerous planets-to-be competed to grow and dominate the swarm of planetesimals. Models show that during the last few hundred million years of planet building there were thousands of collisions between hundreds of protoplanets with sizes comparable to the Moon or even Mars. Collisions stopped only after these bodies coalesced into a dozen or fewer well-separated planets. Any system with a significantly larger number of planets would continue to evolve as long-range gravitational interactions led to intersecting elliptical orbits, collisions, and shattering or merging.

Scars from the last phase of this violent era are preserved on the surfaces of many bodies in the Solar System. The Moon is covered by craters of all sizes, from barely discernible to hundreds of kilometers in diameter (Figure 10.7).

The largest impacts formed the *maria*, which we can see with the naked eye as dark features on the Moon. These basins were formed by such energetic impacts that they fractured the crust to a depth of tens or hundreds of kilometers. Hot magma welled up and flooded the basins with basaltic lava to form the dark maria.

The Moon does not have an atmosphere and its surface suffers far less erosion than features on Earth. The scars produced by ancient impacts have survived virtually unchanged since the solidification of the Moon's crust. Thus, the Moon provides a relatively unmodified record of the impact history of the Solar System since its formation. We can determine the ages of some lunar landforms from laboratory analyses of rock samples returned by landers and the Apollo astronauts. For others, we can estimate relative ages by counting the numbers of craters of various sizes. The oldest regions of the Moon, which are completely covered by craters, are about 4.45 billion years old. The lunar maria, which have much lower crater counts, are a little more than 3 billion years old.

The lunar cratering record indicates a period of heavy bombardment associated with the very last phase of planet building. As the planets reached their final masses, the Solar System teemed with left-over planetesimals. Over the course of the next billion years, this debris battered the surfaces of the planets and created most of the craters we see today on the Moon and on other Solar System bodies. But as they were largely swept up by the larger planets, the population of small bodies in the inner Solar system declined. Compared to our past, the hazard of impacts today, though not zero, is greatly diminished. As peaceful as the full Moon may appear on a summer evening, this belies its dramatic origin in the violent early days of our Solar System.

Pluto and the Kuiper Belt

The outermost planet in our Solar System is Pluto, a tiny object about two-thirds the size of our Moon. It too has a large satellite, Charon, that is about half of Pluto's size. Spectra show that Pluto and Charon are covered by ices, mainly of nitrogen and water, respectively. Discoveries made during the last decade show that there are many more such small icy bodies at the edge of the planetary system outside the orbit of Neptune. Most are kilometer-sized planetesimals resembling giant, dirty snowballs. Pluto may be the biggest of these objects rather than a true planet in its own right. Millions of icy planetesimals, mostly very small, reside in what is now known as the *Kuiper Belt*, discovered in the early 1990s by David Jewitt and Jane Luu. A new bright Kuiper Belt object named Quaoar was discovered in 2002. With a diameter of 1250 kilometers, Quaoar is an exceptionally large icy planetesimal. Many more, even larger objects, must reside in these distant and little known realms. Despite their numbers, the total mass of all the Kuiper Belt objects is likely not to exceed one-tenth of the mass of Earth.

Figure 10.8. Comet Hale-Bopp is one of the most spectacular comets that have appeared in recent years. (Wei-Hao Wang).

The Kuiper Belt extends from the edge of the planetary system to about 50 AU from the Sun. But not all of the Kuiper Belt objects originated in these regions. At least some of them appear to be the remaining members of an initially much larger population of icy planetesimals formed in the protoplanetary disk between Jupiter and Neptune and later scattered out to their present orbits by gentle dynamical encounters with a planet.

Occasional gravitational perturbations can send Kuiper Belt objects back into the inner Solar System. As such objects approach the Sun, their surface layers are heated and evaporate and the dirty snowballs develop a tail. They become visible as *comets*, perhaps the most spectacular objects in the Solar System (Figure 10.8). Comets originating in the Kuiper Belt have short periods of less than 200 years.

Pristine material: comets and the Oort Cloud

Comets with periods longer than 200 years are called long-period comets and are believed to come from the most distant portion of our Solar System. By investigating the orbits and numbers of long-period comets, the Dutch astronomer Jan Oort postulated the existence of a huge roughly spherical reservoir of comets around our planetary system extending to a distance of 50 000–100 000 AU, or about 1 light year. The flux of incoming long-period comets suggests that this *Oort Cloud* contains about ten trillion cometary bodies with radii larger than about 1 km. These comets spend most of their time frozen in deep space in long-period orbits, and take hundreds of thousands or even millions of years to go around the Sun just once. The Oort Cloud is constantly stirred by tidal forces in our Milky Way and occasionally perturbed by the passage of a nearby star or giant molecular cloud. As a result, some comets get thrown into the inner Solar System.

As the Sun warms their icy surfaces, comets eject clouds of glowing plasma and dust which form cometary tails. While the pressure of the Solar wind pushes on the plasma, the pressure of sunlight drives away the grains of liberated dust, thus forming the gossamer tails for which comets are known. Occasionally, when the Earth passes through a stream of dust and larger particles shed by passing comets, we experience a meteor shower (see Chapter 11).

The Oort Cloud is a natural by-product of the formation of the Solar System. The comets of the Oort Cloud were not formed at their present enormous distances from the Sun, but they are believed to be icy planetesimals that originated in the cold region of the giant planets. As the planets grew, their powerful gravitational forces stirred up the remaining ensemble of planetesimals, violently flinging some out of the Solar System to wander among the stars. But not all attained escape speed from the Sun. Some ended up in the Oort Cloud with only the feeblest gravitational bond to the Solar System. The long-period comets provide nearly pristine samples of the building blocks from which the cores of the giant planets were assembled.

The zodiacal light and debris disks around other stars

The space between the planets contains innumerable microscopic dust grains, which are continuously lost from the Solar System. They either spiral into the Sun or are expelled by radiation pressure. Thus, interplanetary dust must be continuously replenished by collisions between bodies in the asteroid belt and loss of dust by outgassing comets as they approach the Sun. From locations with exceptionally dark skies, we can see sunlight reflected by these small particles as a faintly glowing band along the ecliptic, the *zodiacal light* (Figure 10.9). In the past, the Solar System was dustier than today, and the zodiacal light must have been much brighter. We may be able to

Figure 10.9. This image shows the zodiacal light stretching up from the horizon on the right. This faint light comes from sunlight that is reflected off dust grains between the planets. (Wei-Hao Wang).

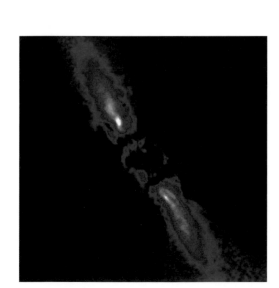

Figure 10.10. The faint glow of a dusty disk around the young star Beta Pictoris is seen in this image, where the bright central star has been occulted by a disk. (ESO).

detect the equivalent of the zodiacal light around other stars if they are sufficiently young.

Such dusty disks have been found around several nearby stars. The most famous example is the southern star Beta Pictoris (Figure 10.10) which is somewhat more massive than the Sun. At an age of about 20 million years, Beta Pictoris is more than 200 times younger than the 4.5 billion year old Sun. It is surrounded by a large edge-on dusty disk which extends more than 1000 AU from the star. The dust in this disk may be produced by collisions

Figure 10.11. A dust disk surrounds the young star AU Microscopii. A planetary system is probably in the process of forming now in the disk. The image is 100 AU wide. The dark pattern obscures the intense light of the central star. (M. Liu/Keck).

of icy and rocky planetesimals and the evaporation of comets. Spectroscopic observations of Beta Pictoris show variable absorption produced by gas falling towards the star, perhaps evidence for decaying giant comets. Curious kinks and density fluctuations in the disk suggest that planets have already formed around Beta Pictoris and are perturbing the disk. Similar *debris disks* have been found around a few other stars such as the 12 million-year-old AU Microscopii where a dusty disk circles the star (Figure 10.11). The disk has small irregularities and clumps, which may be caused by the gravitational effects of shepherding planets.

Debris disks such as those seen around Beta Pictoris and AU Microscopii represent an intermediate stage between newborn T Tauri stars surrounded by massive circumstellar disks and adult stars with fully formed planetary systems. Studies of stars with debris disks may provide insights into the processes that dominated the formation of planets in our own Solar System.

11 Messengers from the past

Pieces of the original building blocks of our planetary system carry with them unique insights into our own early history. We have succeeded in getting material from the Moon, and missions to bring back samples from an asteroid and a comet are planned. But we constantly, and for free, receive material from the space between the planets. Most of this material consists of microscopic dust, and most disappears in the oceans. However, larger pieces can survive their fiery arrival and descend to Earth as *meteorites*.

Finding meteorites in cold and hot deserts

About 50 tons of extraterrestrial material pelt the Earth every day. These pieces can in principle be retrieved, but the difficulty is to distinguish meteorites from ordinary rocks. Identifying a meteorite can be challenging even for experts. However, the potential scientific rewards are so large that many systematic searches for these cosmic intruders have been undertaken. And Earth's deserts have emerged as the place of choice for meteorite hunters.

Over centuries, many thousands of meteorites have been found, catalogued and protected in museums around the world. These range from tiny pebbles weighing only a few grams to gigantic bodies measuring up to 50 tons. Most were found by chance. Only a few were collected after they were seen to fall from the sky. But the real breakthrough in meteorite collecting came in 1969 when Japanese glaciologists on an expedition to Antarctica were puzzled by several small black stones lying on top of clear blue ice (Figure 11.1). Closer examination showed that they were meteorites, and not just pieces from a single body that had broken up upon impact. How could different meteorites fall within the same small region?

Researchers eventually learned that the meteorites did fall in different locations across Antarctica, but were subsequently transported by the movement of the ice and concentrated in this spot. Meteorites that fall on Antarctica are gradually covered by layers of snow, becoming entombed and protected. As the snowpack grows thicker it turns into dense glacier ice which glides slowly towards the outer edges of the continent. Once they reach the ocean, icebergs break off, melt and eventually deposit the meteorites on the ocean floor.

Figure 11.1. Meteorites accumulate on the blue ice fields of Antarctica, and can be collected by specially equipped expeditions. (R.P. Harvey, NASA/NSF).

Where the advance of the iceflow is blocked by coastal mountains, the meteorites are forced to the surface. As strong winds and sunshine sublimate the glaciers, their entombed load of meteorites is left behind on the surface. Over thousands of years, meteorites are concentrated on clean ice surfaces behind mountains. These dark rocks are easily visible in stark contrast to the blue glacier ice, and they can be picked up by researchers who brave the mind-numbing cold, snow squalls and isolation.

In the past 30 years, more than 15 000 specimens have been collected in Antarctica's blue ice fields. Many meteorites break up into several pieces either when they hit the Earth's atmosphere or while they are under the pressure of the ice layers. Therefore this collection represents perhaps 3000 or more distinct meteorite parent bodies. This more than doubles the number of meteorites collected over the past few hundred years all over the world.

Antarctica is very dry, with low precipitation, and is a *cold* desert. More recently, it has been realized that *hot* deserts, like those in north Africa, west Australia and Chile, are also treasure chests full of meteorites. Water

Figure 11.2. A meteorite in the Chilean Atacama desert has rested in this location for 6000 years until found by co-author Bo Reipurth.

seeping into cracks is one of the principal enemies of these rocks from space, eventually making them crumble into powder. Dry climates preserve meteorites better, and so a growing effort has been devoted to searching some of the most inhospitable deserts of the world for meteorites (Figure 11.2).

Types of meteorites

Analysis of the many meteorites collected over the years reveals that they can be divided into several broad categories. The best known and most easily recognized are the *iron* meteorites which consist of nickel-iron metal alloys. Because most people take notice of a piece of metal on the ground, collections contain a disproportionate amount of iron meteorites. In reality, only about 5 percent of those meteorites that are observed to fall to Earth are irons. An even smaller fraction (about 1 percent) are *stony-iron* meteorites which consist of a mix of iron, nickel and silicates.

By far the most common visitors from space are the *stony* meteorites, which resemble rocks found on Earth in their density and appearance. Almost all stony meteorites are *chondrites*, named so because they contain numerous millimeter-sized spherical inclusions, called chondrules,[1] embedded in a rocky matrix (Figure 11.3). Chondrules are not found in terrestrial rocks, so their presence in a sample is proof of its cosmic origins. A rare and extremely interesting subclass of chondrites are the *carbonaceous chondrites*. As their name suggests, they contain carbon together with a variety of other complex organic molecules. They are also unusual in that they contain water bound to their minerals; in some cases they consist of as much as 20 percent water. Most carbonaceous chondrites are very friable, and if not recovered shortly after landing they will turn into a gray powder. Finally, there is a minor group of stony meteorites which do not have chondrules called *achondrites*.

Meteorites as interplanetary flotsam

The majority of meteorites have ages of about 4.5 billion years, determined from the analysis of their now extinct radioactivity.[2] Consequently this is the age astronomers have come to accept for our Solar System. So meteorites are messengers from our distant past, bringing us information about conditions at the time when the planets were built. But meteorites have been altered through the eons. Analysis suggests that all meteorites except the chondrites have been melted and transformed. The various categories of meteorites can help us piece together how our planetary system formed and evolved.

Meteoriticists agree that carbonaceous chondrites are among the most primitive, almost unprocessed, material dating from the time the planets formed. Inside chondrites one can occasionally find interstellar dust grains which miraculously survived from the molecular cloud out of which the Sun and the early Solar nebula accumulated. The chondrules found in chondrites were at one time droplets of molten rock floating around the early Solar nebula. They were heated and cooled very rapidly, preserving their spherical shapes. The formation of chondrules is one of the mysteries of early Solar System research. Enormous lightning discharges in the early Solar nebula are one possibility. Heating by giant flares on the proto-Sun or by powerful

Figure 11.3. Chondritic meteorites show chondrules when cut open. Polarized light image of ALH78119 from Antarctica. (Dante Lauretta).

shock waves are other possibilities. Molten spray from impacts between planetesimals is yet another. Whatever their origin, chondrules must have gradually stuck to other solids in the Solar nebula to eventually form chondritic material. Most chondrites have been well protected since their origin. A likely place to hide was in the upper layers of planetesimals, buried under kilometers of debris. Major collisions later released some of this chondritic material, which eventually found its way to Earth.

The achondrites, stony meteorites without chondrules, resemble terrestrial lava rocks. They may have been chondritic material in planetesimals which suffered impact melting. Consequently their top layers melted and briefly flowed as lava streams before solidifying. Like the deeper chondritic material, achondrites would eventually be ejected into space by subsequent collisions.

As the planetesimals grew in size, intense radioactive decay, inherited from dying stars by the young Solar System,[3] heated their interiors. But it can take a long time for heat to escape from the interior of a planetesimal because of the insulation provided by the thick layers above. The interiors of the larger planetesimals therefore melted, and heavier elements like iron and nickel sank, forming solid metal cores. As the Solar System aged, radioactivity declined and planetesimals solidified. Iron meteorites most likely come from such deep planetesimal interiors and were uncovered by catastrophic collisions. And the rare stony-iron meteorites may possibly come from the interface regions between the upper stony layers and the iron cores.

Parent bodies and left-over planetesimals

Most meteorites appear to have come from the main asteroid belt located between the orbits of Mars and Jupiter, where more than 100 000 minor planets with sizes from a few to a thousand kilometers have been identified. But what is the origin of the asteroids? It is believed that asteroids are the last planetesimals, the building blocks of the planets. Or perhaps they are the *remnants* of planetesimals, because in the 4.5 billion years since their formation, they have been heavily pummeled by encounters with orbiting debris. When two comparably sized planetesimals collide, serious damage and even catastrophic breakup can result. The asteroids are thus the battered remains of the planetesimals left over from the earliest days of the Solar System. Close-up photographs of asteroids taken by spacecraft testify to their violent histories (Figure 11.4). Most meteorites are the splinters ejected by these dramatic encounters.

But why did these planetesimals not accrete into yet another planet? Detailed computer simulations suggest that the proto-Jupiter may have inhibited the formation of another planet in the asteroid belt. The powerful gravity of Jupiter, even before it accumulated all of its bulk, could have stirred the planetesimals, flinging some into orbits that would cross the fledgling interior planets. Others were thrown out of the Solar System. The remaining ones were left so perturbed that instead of coalescing, the planetesimals would fragment and disrupt when they experienced an encounter. Jupiter may have prevented another planet from forming in the present-day asteroid belt.

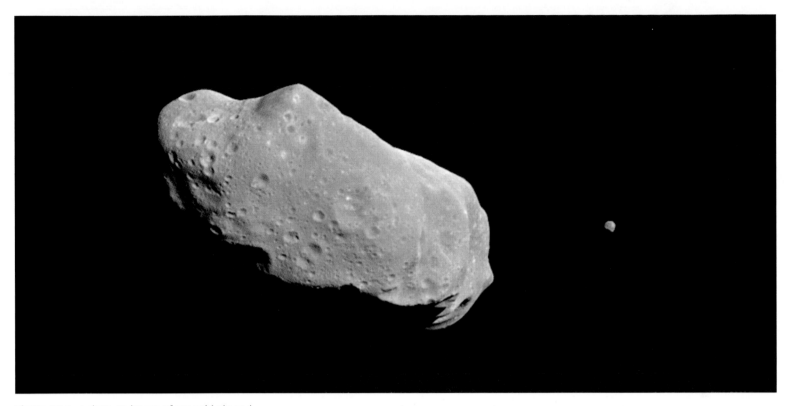

Figure 11.4. A close-up image of asteroid Ida and its tiny moon Dactyl obtained with the Galileo spacecraft. (NASA/JPL/Galileo).

As discussed in the previous chapter, there is another reservoir of planetesimals in the present-day Solar System, namely the Kuiper Belt, located beyond the orbit of Neptune. In the permanent deep-freeze of these distant realms, planetesimals formed mostly out of snowflakes mixed with dust, resulting in bodies resembling giant kilometer-sized dirty snowballs. It is estimated that 100 million of such icy planetesimals larger than 1 km reside in the Kuiper Belt. Occasionally gravitational perturbations send these bodies into the inner Solar System to become comets.

Each time a comet swings around the Sun it loses mass and eventually completely evaporates. But the dust grains and rocks released from the snowy matrix of a comet are scattered along its orbit. When the Earth crosses an old comet orbit, we are treated to a meteor shower. During the best annual meteor showers hundreds of meteors can be seen each hour at their peak which typically lasts only a few hours. In rare instances meteor storms occur with tens of thousands of shooting stars lighting up the sky every hour (Figure 11.5). The last great meteor storm occurred in November 1966. However, the Leonid storms of 2000–2 also produced notable showers of several thousand meteors per hour.

Probably most of the dust and small pebbles embedded in the heads of comets are similar to friable carbonaceous chondrites and so are unlikely to survive the fiery passage through the Earth's atmosphere. Soon space missions will recover samples from a comet and acquire pristine material from the frozen planetesimals at the edge of the Solar System.

Figure 11.5. Every year the Earth crosses the orbits of ancient comets and encounters numerous dust particles. On rare occasions a meteor shower results, as in this photo of a burst of Leonid meteors. (Tony and Daphne Hallas).

Figure 11.6. The heavily cratered lunar surface including the 93 km diameter crater Copernicus, as seen in an image taken from Apollo 17. (NASA).

Bull's-eyes and near misses

We tend to think of meteorites as fist-sized rocks or smaller. Indeed, the vast majority of meteorites are very small, although occasionally larger bodies penetrate the Earth's atmosphere. This is, however, so rare that historical records are of little help when we try to determine the impact rate of larger bodies. But the lack of an atmosphere and active geology on our Moon makes it ideal for estimating impact rates. On the Moon all craters are preserved until obliterated by other impacts (Figure 11.6). By counting craters of different sizes on the lunar surface and estimating their ages, astronomers have estimated the influx of larger bodies in the neighborhood of the Earth. Impacts of meter size bodies occur several times every year. Bodies with a diameter of 100 meters hit the Earth once every thousand years. Kilometer sized bodies crash into the Earth at intervals of a

Figure 11.7. The Wolfe Creek structure in Australia is an astrobleme formed by the impact of a giant meteorite about 300,000 years ago. (L. Stougaard Nielsen and K. Bucka-Lassen).

million years. Although such giant impacts are rare, they happen and can be devastating.

Scattered around the Earth, geologists have found at least 15 huge circular structures more than 30 km in diameter which are meteoritic in origin. These *astroblemes* have ages between 40 million and almost 2 billion years (Figure 11.7). Many more have been eradicated by erosion, weather, oceans, and geological processes. The most famous of all impacts occurred 65 million years ago and caused the extinction of the dinosaurs.

Even in historic times violent impacts have occurred. In the early morning of June 30, 1908 a body estimated to have had a diameter between 30 and 60 meters exploded six kilometers above the forests of Siberia. Four hundred kilometers away, observers reported seeing a ball of fire and hearing deafening thunder. At a trading station 60 km away, a burning wind blew out windows and the ground trembled. Some herders camping 30 km away were awakened as their tent was blown into the air. An old man who was thrown 15 meters before hitting a tree died from his wounds. The heat was so intense that the forest was set ablaze. Because the object exploded in the atmosphere, it left no crater. But up to 15 km away, trees were blown down, radially pointing away from ground zero.

Fortunately, Siberia in 1908 was among the most sparsely populated regions on Earth. But disaster nearly struck the USA in 1972. On April 10 of that year, a body estimated to be the size of a house entered the Earth's atmosphere above the western states with a speed of 15 km per second. By luck it came in at such a low angle that it flew as a brilliant fireball across Utah, Idaho, and Montana without getting closer to the ground than about 50 km. It was seen in plain daylight by thousands of people and photographed and filmed. Finally it left the atmosphere somewhere over Canada. It is now circling the Sun as a charred and blackened body. Some day in the future it will be back.

For several decades, defense satellite networks have monitored the Earth for nuclear explosions and hostile rocket launches. Each year their sensors record explosions in the atmosphere caused by large meteorites consisting of very fragile and loose material. Such explosions often release the energy of a modest nuclear warhead.

Several groups of astronomers have undertaken systematic searches of space near the Earth to get a more accurate idea of how often the Earth may encounter giant bodies. The most advanced of these projects is the so-called LINEAR program. Every clear night astronomers use special instruments and software programs to automatically image the sky and find the positions of objects that move rapidly across the sky, an indication that they are close to the Earth. Large numbers of such "Earth crossing" objects have already been found. As the data from this and future programs gradually build up, we will better understand the risks we face from the sky.

12 Hazards to planet formation

The Great Nebula in Orion (introduced in Chapter 9) is the best studied region of star formation in the sky (Figure 12.1). Only a few light years in diameter, this stellar nursery has forged several thousand stars over the course of the last million years (Figures 12.2 and 12.3). At least a half dozen are massive and therefore luminous. Lit by the dazzling light of these O and B stars, the Orion Nebula is bathed in intense radiation and churned by violent stellar winds. Gossamer clouds of glowing hydrogen and shimmering sheets of reflecting dust shroud the region in a veil of diffuse light. The Orion Nebula provides many important clues about the birth of stars, the evolution of circumstellar disks, and the hazards faced by nascent planetary systems.

Young stars and disks in Orion

In 1993, the Hubble Space Telescope obtained images of the Orion Nebula with an extremely high resolution of about 0.1 arcsecond. On this scale, many Orion Nebula stars were found to be surrounded by small comet-shaped nebulae of glowing gas pointing directly away from the most massive star in the region (Figure 12.4). A study of these images led by C. Robert O'Dell reported the discovery of circumstellar disks surrounding low-mass stars in the Orion Nebula. This study explained the small luminous structures as interaction zones between dense disks and their environments (Figure 12.5). This seminal paper presented the first direct images of protoplanetary disks around young stars ever obtained.

The term "proplyd" (an abbreviation of the phrase "PROto-PLanetarY-Disk") was coined to describe these teardrop-shaped plasma clouds and their embedded disks. No longer were disks confined to the imagination of inventive theorists; they could be seen directly on high-resolution images!

Unfortunately, the quality of the 1993 images was impaired by the misshapen figure of the primary mirror on the HST. However, during the first Space Shuttle repair mission in 1994, a new camera was installed on Hubble with optics to correct the aberrations introduced by the telescope's faulty mirror. New and sharper images of the Orion Nebula revealed dozens of new disks surrounding low-mass stars (Figure 12.6). By now, about 500 of the Orion Nebula's young stars have been carefully inspected by HST. More than 200 are surrounded by tear-drop shaped proplyds. Disks are directly

Figure 12.1. The region south of Orion's Belt includes the large Orion Nebula, the Horsehead Nebula, and large cloud structures. (M. Bissell & R.S. Sutherland).

Figure 12.2. The bright central part of the Orion Nebula as seen at optical wavelengths. (R. O'Dell, NASA/STScI).

seen within at least 60 bright proplyds. Additionally, 20 bare disks not surrounded by ionized envelopes are detected in silhouette against the nebula. These "dark proplyds," or "silhouette disks" must lie in the foreground sufficiently far from Orion's massive stars to avoid being ionized.

Earlier ground-based observations hinted at the existence of these proplyds. Photographic and electronic images taken in the 1970s under the best seeing conditions with large telescopes revealed clouds of extended hydrogen around several low-mass stars in the core of the Nebula. About a decade later, the *Very Large Array* radio telescope in New Mexico found that some of the Orion Nebula's young stars were surrounded by crescent-shaped clouds of plasma facing the nebula's most massive and luminous star. HST found disks surrounded by bright crescents of hydrogen emission at exactly the location detected by the VLA a decade earlier.

The early suggestions of extended structures surrounding some of Orion's T Tauri stars were puzzling. Given the angular diameters of these clouds (about a half arcsecond or 250 AU), the gravitational pull of their central

Figure 12.3. The inner part of the Orion Nebula as seen at infrared wavelengths. (M. McCaughrean, ESO).

stars is insufficient to hold 10 000 degree plasma in place. The clouds had to be expanding because the typical speeds of ions far exceed the star's gravity at that distance. In the 1990s, spectroscopic observations demonstrated that hydrogen gas was streaming away from these objects. To persist for more than a century, the gas lost from these clouds had to be replenished from an unseen, but very dense reservoir. The reservoirs could not be spherical since dust would obscure the central stars. On these grounds, it was predicted that disks should surround a number of the stars observed in Orion. And this was indeed what HST found when it studied the Orion Nebula region with its superior resolution.

Evaporating disks

Why do proplyds have extended envelopes and cometary shapes with tails pointing away from the source of light? Because ultraviolet radiation heats the surfaces of circumstellar disks. In the outer portions of disks where gravity binds material only weakly to the parent star, the UV-heated gas can escape into the surrounding HII region. For example, particles located

Figure 12.5. Two disks embedded within teardrop shaped plasma clouds (HST 10 and 17) and a nearly face-on disk seen only in silhouette (HST 16). The central star is visible in the center of the face-on disk while the stars in the edge-on disks are completely obscured. (J. Bally, NASA/STScI).

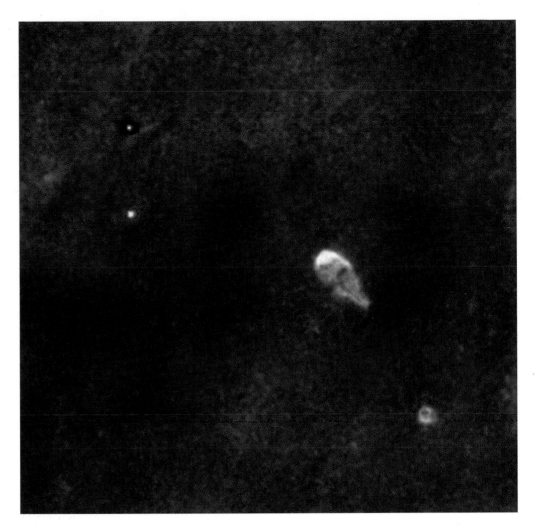

100 AU from a one solar mass star have orbital speeds of about 3 km/s. If atoms in this region are accelerated by an additional 1.5 km/s, they can escape.

The effect of UV radiation depends on the wavelength. Energetic "hard" UV with wavelengths less than 912 angstroms ionizes hydrogen.[1] The lower-energy "soft" UV radiation with wavelengths longer than 912 angstroms leaves hydrogen atoms intact. However, soft-UV light can dissociate molecules and ionize trace elements such as carbon. But the most important effect is heating. While soft-UV light heats gas to a few thousand degrees, hard-UV radiation increases the temperature to nearly 10 000 K. Irradiated gas becomes warm, expands, and escapes the disk in regions where its speed is larger than the gravitational escape speed.

As it expands, the flow becomes less dense, allowing hard-UV radiation to penetrate and ionize the gas. At several disk radii the wind therefore flows through an "ionization front" where the temperature jumps to nearly 10 000 K and the plasma accelerates to more than ten kilometers per second. Circumstellar disks, their winds, and ionization fronts cast shadows. Thus,

Figure 12.6. A single proplyd floats in the Orion HII region. The star is surrounded by a dark silhouette disk. (J. Bally, NASA/STScI).

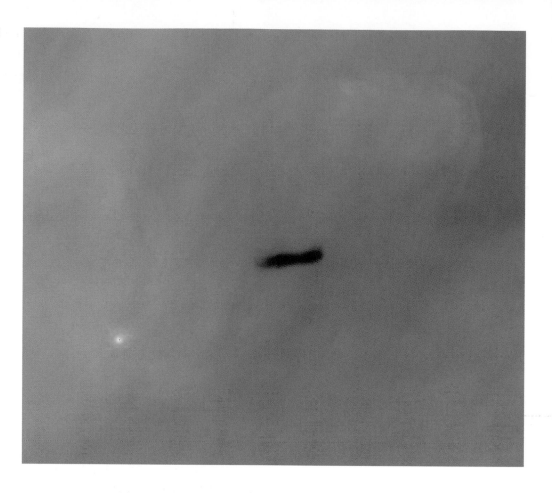

Figure 12.7. Two proplyds that both show small collimated jets emanating from their central stars. (J. Bally, NASA/STScI/HST).

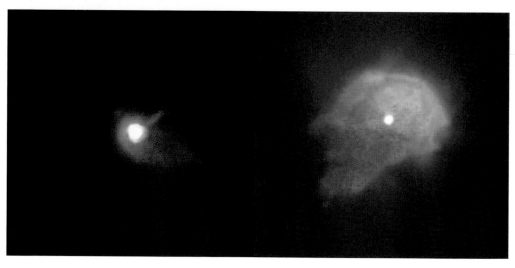

gas in the shadow is shielded from direct illumination and forms the proplyd tails.[2]

Two proplyds with large ionization fronts are shown in Figure 12.7. Some young embedded stars power high-velocity jets which produce the spikes of emission emerging from these proplyds.

Orion's large proplyds lose nearly a millionth of a solar mass per year. If this mass-loss rate remained steady, a disk which started with 1 percent of the Sun's mass, about the minimum required to form a planetary system like ours, would disappear completely in only 10 000 years. But as disks shrink, their mass-loss rates decline. It can take more than a million years for a disk to shrivel to an outer radius of 10 AU. Thus, small disks may survive long enough to form planets.

As the disk outer radius shrinks to a about 40 AU (for a one solar mass star), the soft-UV heated gas can no longer leave the system. With no wind to absorb the hard-UV radiation, the ionization front descends to the disk surface. Mass-loss resumes as a fully ionized wind erodes the disk until its outer radius is less than about 4 AU. However, such small disks will be indistinguishable from ordinary stars even in HST images, since disks with radii less than 40 AU tend to be lost in the glare of their central stars.

About half of the Orion Nebula's stars are embedded in large proplyds experiencing soft-UV dominated mass-loss. But excess infrared emission indicates that most of the remaining stars are also surrounded by disks which are too small to be seen in images and which may be in the hard-UV dominated phase of mass-loss.

While massive stars are the principal sources of UV radiation, low-mass stars also produce some UV light. The UV luminosity of a typical T Tauri star is more than a million times lower than the UV output of a massive star. Yet, this radiation has a relatively large impact on disk evolution. While massive stars erode disks from the outside, self-irradiation erodes disks from the inside. Self-irradiation by hard-UV drives mass-loss from an annular zone 5–10 AU from the central star. Closer to the star, gravity can hold on to the hard-UV irradiated plasma. The wind rising from this annulus scatters stellar UV light and emits its own soft-UV glow. This radiation shines onto the outer parts of the disk where it can drive mass-loss from beyond about 40 AU. Thus, disks erode from both outside and inside.

Viscosity in a disk spreads matter radially.[3] The disk's outer edge tends to expand to larger radii, gas at an inner edge tends to migrate to smaller radii, and gaps tend to fill in. Thus viscosity feeds gas into the zones where UV radiation can remove it.

UV from nearby massive stars and self-irradiation from the central T Tauri stars can remove all hydrogen and small dust particles on a time-scale of a few million years. If planetary systems are abundant, planetesimals and hydrogen-rich giant planets must form in less than a few million years.

Clusters and collisions

Large disks face another hazard in the star-forming environment because most stars form in groups or clusters. As stars move about in response to the cluster's gravitational pull, they occasionally suffer close encounters. During a close stellar passage, disks can crash into each other and be dispersed.

Clusters can have extremely high densities during their formation. The central density of the Orion Nebula cluster is about 3000 stars per cubic light year: nearly ten million times higher than the density of stars near the Sun. While the typical distance between stars in the solar neighborhood is about three light years, the average spacing between stars in the core of the Orion Nebula is only a few hundredths of a light-year.

Although star formation has by now stopped in the Orion Nebula, it continues in the OMC1 cloud core, located less than a light year behind the rear ionization front of the Orion Nebula. In OMC1, we see several high-mass stars separated from each other by no more than about a thousand AU. We do not yet have the capability to see dimmer, lower mass stars due to heavy obscuration and confusion. When the expected numbers of these missing low-mass stars are included, the density of stars may well be greater than 30 000 stars per cubic light year. With stars packed so close to each other, there is not much room for circumstellar disks. The situation is similar to that in multiple stars. The mean spacing between individual members of multiple star systems is typically less than a few hundred AU, not much larger than the dimensions of circumstellar disks inside Orion's proplyds.

Large cloud cores like OMC1 produce veritable swarms of smaller cores, disks, and protostars, which for a brief period (less than 100 000 years) can interact violently. Cores, disks, and stars swarm around in a frenzied cosmic dance. Mutual gravitational tugs lead to near misses, the formation of some capture-formed binaries, the violent expulsion of a few stars, and perhaps even to mergers. Models show that the outer radii of surviving disks will be chopped down to values comparable to the closest separation of the stars during encounters.

As two stars swing past each other, their disks become distorted by tidal forces. The outer parts of disks may be accelerated and tossed out of the system, much like a rock hurled by a slingshot. Models of interactions show that the close passage of an intruding star can generate strong spiral patterns, ejected tails of disk debris, and chaotic motions in disks. Only the very tightly bound inner disks survive. The chaos and violence of mutual interactions can destroy protoplanetary disks and eject previously formed bodies into the interstellar environment. Dynamical ejection is one way in which free-floating planets and brown dwarfs are produced (see Chapters 5 and 13).

If a disk does manage to survive long enough to produce a planetary system, the system may be disrupted by sibling stars if the parent cluster remains bound by its own gravity and survives as an open or globular cluster. The Hubble Space Telescope searched for giant planets in two globular clusters, Messier 22 and 47 Tucana, looking for the minuscule fading of light when a planet transits the disk of the star it orbits (see Chapter 13). These experiments proved negative: no planets were found. In hindsight, the outcome is not surprising. During its formation, these clusters must have contained dozens or perhaps even hundreds of massive stars. Intense UV light would have photo-ablated young disks. Over the billions of years that such clusters live, continued interactions with sibling stars produce an additional hazard. Even if planets did manage to form, there is a good chance that they were ejected from their parent systems.

Supernovae and disks

OB associations spawn dozens of massive stars which end their lives in supernova explosions. All massive stars die within 40 million years of their birth, comparable to the time-scale on which rocky planets are assembled by collisional build-up. Most young stars in an expanding association will still be near their birthsites when the first massive stars die.

During the last stages of stellar evolution culminating in a supernova explosion, many massive stars oscillate between red and blue supergiant phases. For a period lasting hundreds of thousands of years, the UV luminosity of blue supergiants can reach 10 million Suns, more than 100 times the luminosity of Orion's Trapezium stars. These supergiants eventually explode. For several weeks to months, the UV light generated by a supernova can be millions of times greater than that of the Trapezium. The huge UV luminosity of supergiants and of supernovae can greatly enhance the photo-erosion and mass-loss rates of nearby clouds and proplyds.

A supernova's expanding debris can contain as much energy as the Sun generates in its entire lifetime. The remnant sweeps up the surrounding HII region and blows past protostars and their disks. The passage of a supernova blast wave is akin to a powerful but brief hurricane that blows past stars, disks, and star forming cloud cores. Weakly bound outer disks can lose a bit of additional mass over and above that lost to UV irradiation. However, the supernova blast waves are not likely to disrupt surviving disks.

A passing supernova blast wave, on the other hand, can contaminate protostellar clouds and disks with the freshly synthesized elements and radioactive isotopes produced during the massive star's explosive demise. These atoms work their way into grains of interstellar and protoplanetary dust. As they coagulate to form planetesimals and planets, the heating produced by the decay of radioactive elements liquifies their interiors. Thus, in solid bodies larger than about 100 kilometers in diameter, heavy materials such as nickel and iron settle to the center, and light materials such as silicate minerals rise to the surface.

Our own Solar System may have been run over by the blast wave from a supernova. Analysis of meteorites has revealed the daughter products of short-lived radioisotopes.[4] This may be evidence for contamination of protoplanetary material by a nearby supernova which exploded within a few million years of the formation of our own Solar System.

Can planetary systems form in proplyds?

It takes millions of years to build planets in the "standard model" of planet formation (Chapter 10). Yet, Orion's proplyds are observed to be losing mass at such fast rates that most shrink to invisibility in less than a million years. The central issue to consider is the time-scale in which mass-loss occurs from the inner parts of a circumstellar disk due to both external and self-illumination.

The success or failure of planet formation in environments such as Orion has profound implications for the existence of planets around all stars. The majority of the stars in the sky form in OB associations and are irradiated by intense UV radiation. If planets can form around stars whose disks are evaporated soon after birth, most stars may come to possess planetary systems. On the other hand, if UV light and stellar encounters prevent planet formation, most stars may lack planets. UV radiation may mostly impact the formation of gas giants such as Jupiter which consist primarily of hydrogen and helium, the gases most readily stripped from a disk.

To form a gas giant in the standard model, a large, rocky, proto-planetary core accumulates from the collision of smaller planetesimals, as discussed in Chapter 10. Once a core reaches a mass of five to ten times that of Earth, its gravitational pull can accrete light gases such as hydrogen and helium from the surrounding disk. In an Orion-like environment, light gases from the outer disk are rapidly stripped away by UV radiation. Self-irradiation may strip these species from beyond 5 AU in a few million years. If proto-planetary cores take longer than a few million years to form, there may not be any gas left to accrete. Therefore, the formation of giant planets must occur either in less than a few million years, or gas giants may not form in UV-rich environments.

Giant proto-planets have radii as large as an AU while accreting gas from their parent disks. Proto-planets formed in the outer disk are exposed to UV radiation and therefore suffer mass-loss. But a 1 AU diameter object can survive for tens of millions of years in an irradiated environment such as Orion. Mild depletion of hydrogen will leave behind intact gas giants like Saturn or Jupiter that have compositions fairly close to that of the Sun. Severe depletion of hydrogen, on the other hand, will result in worlds such as Uranus and Neptune, which have much larger contents of heavier elements.

In Orion, the outer radii of disks shrink to where they can no longer sustain soft-UV driven mass-loss in a few hundred thousand years. But it takes many millions of years for the disk to evolve to the end-point of its ionization-dominated mass-loss phase. Although viscous migration of matter from the inner disk to its outer edge may shorten disk lifetimes, complete disk dissipation takes millions of years. Even in harsh, radiation-rich environments, the inner parts of disks within 10 AU may survive long enough to form planetary systems.

Radiation fields may actually *accelerate* the early phases of planet formation. Large dust grains and ice particles are left behind by UV-induced mass-loss. These particles contain the heavier elements. In disks which have had sufficient time to grow particles larger than a few hundred microns prior to UV exposure, disk evaporation will selectively increase the abundance of heavy elements in the disk. As discussed in Chapter 10, solids with sizes ranging from centimeters to meters settle towards the disk mid-plane. These particles tend to orbit the central star at the local Kepler speed while the gas orbits slightly slower. Thus, rocks feel a head-wind and will therefore drift towards the central star. When in-spiraling solids encounter gas-free gaps in the disk (such as the inner disk gap produced by the stellar magnetic field, or gaps created further out by rapidly forming planets) the radial drift of solids stops and the particles will pile up, increasing the likelihood of a planet forming there.

The combined impact of these four processes – the removal of hydrogen and other gas-phase material, grain growth and sedimentation toward the disk mid-plane, radial drift of large solid particles and ice, and the halting of migrating solids just outside gaps – will concentrate large "metal"-rich solids at certain disk locations. Under the influence of their own self-gravity, kilometer-sized planetesimals can condense out of the disk in only a few hundred years. In this way UV radiation may indeed accelerate the onset of planet formation.

Towards a variety of planetary systems

As we have seen, the environment of a young star depends strongly on the proximity of massive OB stars and on the density of neighboring stars. The large range of environments in which planetary systems form and evolve can produce a rich diversity of system architectures.

If a very young and relatively low-mass disk in which solids have not yet had time to grow becomes exposed to intense UV radiation and loses most of its hydrogen before large rocky cores of giant planets form, the resulting planetary system may never come to possess gas giants such as Jupiter. But if such a disk has sufficient time to grow large protoplanets *before* hydrogen is lost, gaseous giant planets might form by the "standard" core-accretion mechanism. And if the disk forms with enough mass to allow the onset of gravitational instabilities, many gas giants may form very fast.

The early evolution of a system of planets depends on the environment. In particular, planets do not always remain in their original orbits. The birth of gas giants may trigger the formation of spiral waves in the remaining disk which can transport orbital angular momentum efficiently. The inner planets in such a disk may be doomed to fall into their parent stars while the outer planets may drift to larger orbital radii.[5] But if the gas disk that survives past the era of planet formation can be dispersed rapidly, orbital migration can be halted. Some giant planets may migrate inward only to be permanently parked in close-in orbits when the remnant disk is dispersed. If too many planets are produced from a massive disk and parked in semi-permanent orbits, mutual gravitational interactions between them will sooner or later eject some members. Computer simulations show that in systems containing a dozen or more planets, some will be ejected by mutual gravitational interactions.

In Orion-like regions, most low-mass stars started forming well before the massive stars brought accretion to a halt and began photo-evaporating disks. Dust grains may have had over a million years to accumulate mantles of ice and to coagulate into rocks and dirty snowballs with diameters of centimeters or more. Even kilometer-sized planetesimals may have formed well before the onset of UV irradiation (see Chapter 10). Large solid bodies are relatively immune to destruction by UV radiation. If every incident UV photon knocks one atom off an exposed solid body 1 meter in diameter, it can last for more than a million years at three light years from a typical massive star. Consequently, asteroids, comets, and rocky planets can survive in UV-rich environments.

Disks in dark clouds without massive stars may not experience photo-ablation of their outer disks. The outer parts of disks in stars born in isolation may survive intact for much longer than in a UV-rich environment. This gas may accumulate onto giant planets and comet-like objects. However, the inner portions of such disks may still suffer extensive self-irradiation. We do not yet fully understand the different possible outcomes of disk evolution in dark clouds.

Did our own Solar System form in an isolated dark cloud core such as in Taurus, in an OB association such as Orion, or in an even more exotic environment? We cannot tell from our stellar neighbors or Galactic location. If our system originated in a rich cluster or Orion-like environment, our low-mass sibling stars would have completely dispersed and drifted away to the far corners of the Milky Way during the last 4.5 billion years. Any massive stars would have died a long time ago. The best hope for clues comes from meteorites and comets. The meteorite record provides some evidence for birth in an OB association. The presence of the daughter products of short-lived elements and isotopes indicates the injection of supernova debris into materials that formed the Solar System. Some meteorites contain minute amounts of dust particles derived from the atmosphere of a red giant star. There is growing evidence that the Sun and Solar System formed in a complex environment with multiple sources of freshly synthesized isotopes and elements, likely an ancient Orion-like OB association. The return to Earth of comet samples may soon provide additional clues about the type of environment in which our Solar System formed.

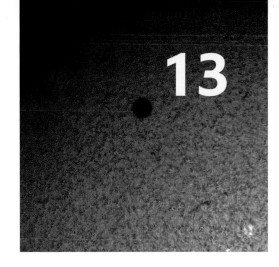

13 Planets around other stars

Is our Solar System unique? Are there planets similar to Earth in orbit around other stars? Our growing understanding of star birth and the recent indirect detections of over 100 giant planets orbiting distant stars (known as "extra-solar" planets or "exo-planets") are starting to provide answers.

The quest for other worlds

As discussed in previous chapters, dense accretion disks (the type of environment in which astronomers expect that planets form) have been detected around many forming and young stars. Their masses and sizes are about right to produce planetary systems similar to our own Solar System. Though the conditions required for planet formation appear to be common, it is by no means clear that planets will inevitably form out of such disks. As we saw in Chapter 12, stellar birth is a violent process and planet formation may be frequently hindered. Only direct observations will determine if planets are common among the stars in our Galaxy.

Planets around other stars are extremely difficult to detect. At visible wavelengths, planets shine by reflecting the light of the stars they orbit. A typical terrestrial (Earth-like) planet intercepts at most about one part in a *billion* of the light of its parent star. Thus, such a planet will appear more than a *billion* times fainter than the star it orbits! Even giant planets such as Jupiter, which can reflect hundreds of times more light because of their larger surface areas, will be extremely faint. Furthermore, the typical separation between planets and their parent stars is so small compared to their distances from us that the feeble planetary light tends to be lost in the star's glare. It is not surprising that no telescope has yet directly seen the light from a planet outside our Solar System. Instead, successful detections of external planets have employed indirect methods.

Giant planets and wobbling stars

In late 1995 and early 1996, two groups of astronomers announced the discovery of giant planets orbiting nearby stars. Michel Mayor and Didier Queloz found a planet with a mass comparable to Jupiter around the star 51 Pegasi. Soon after, the team of Paul Butler and Geoff Marcy announced the discovery of several others.[1] For years, these teams monitored the spectra of Sun-like stars, looking for the subtle periodic motion that an orbiting planet would induce.

The most massive planet in our Solar System, Jupiter, orbits the Sun every 12 years with a speed of 13 km/s, nearly twice the speed at which the Space Shuttle orbits the Earth. As the Sun's gravity pulls on Jupiter, it moves in its orbit around the Sun. But Jupiter also pulls on the Sun. Thus, both objects orbit their center of mass. Since the Sun is a thousand times more massive, it responds to Jupiter's gravitational pull by orbiting the center of mass with a speed of only 12.5 meters per second (28 miles per hour), comparable to the speed of a car driving in a city. The next most massive planet, Saturn, induces a Solar motion of only about 2.7 meters per second.

In the same way, the gravitational pull of a massive exo-planet induces a *reflex motion* in the parent star it orbits. The star's motion alters the observed wavelengths of the light emitted or absorbed in its outer layers: this is the Doppler effect mentioned earlier. Though these Doppler shifts are extremely small, they can nevertheless be measured for distant stars with spectrographs mounted on large telescopes.

The search for reflex motion requires great patience. To detect a Jupiter-like planet orbiting a distant Sun-like star, an astronomer has to make repeated and precise measurements of stellar Doppler shifts over a large part of an orbit, which for the Sun and Jupiter takes 12 years. This spectroscopic method has been used to search for planets around several thousand of the brightest single Sun-like stars. Over 100 distant giant planet candidates have been found by the reflex motion of their wobbling parent stars.

Most searches for exo-planets have concentrated on Sun-like stars. More massive stars have too few spectral lines that can be used for accurate radial velocity measurements. Furthermore, these stars tend to spin faster than the Sun, resulting in broader spectral features which additionally limit the accuracy of radial velocity determinations. Stars much less massive than the Sun tend to be very dim, making precise Doppler measurements difficult.

Much to the surprise of astronomers, most of these distant planetary systems are very different from our own Solar System. The star 51 Pegasi is orbited by a planet about half the mass of Jupiter every 4.23 *days*.[2] This implies that the planet is located much closer to 51 Peg than Mercury is to the Sun. Such an extreme proximity to the parent star results in a furnace-like temperature of over a 1000 degrees centigrade for its outer layers. All planets discovered so far by the spectroscopic reflex motion method have masses comparable to Jupiter and tend to be in relatively short period orbits. These planetary systems are dominated by giant worlds moving along small orbits similar to those of the Earth-like planets in our own Solar System. Apparently, Nature can produce planetary systems with architectures radically different from our own Solar System. It appears that at least 5 percent of all Sun-like stars are orbited by close-in giant planets. Because of the proximity to their parent stars, these objects have been named "hot Jupiters."

While most extra-Solar gas giants are in very short-period orbits, a few have been found with orbital periods of many years, indicating that they are in orbits similar to our Jupiter. Several systems have been found which contain two or even three giant planets.

That we have only detected gas giants around mostly single Sun-like stars is, to a large measure, a selection effect that reflects the biases of the detection method. Close-in gas giants produce the most easy to detect radial velocity signal with a relatively large amplitude and short period. The same planets in larger orbits would produce longer periods and lower amplitude radial velocity signatures which are harder to detect. A hypothetical astronomer living on a distant planet and studying our own Solar System would find it very hard to detect Jupiter and Saturn with our current radial velocity measuring instruments. And terrestrial planets produce signatures that are orders of magnitude below what can be currently detected. For the discovery of terrestrial planets we require different approaches.

Planets in silhouette

In our own Solar System, the inner planets Mercury and Venus are occasionally seen to move across the face of the Sun in an event called a planetary *transit*.

Transits were important to geographers several centuries ago. They provided one of the few means by which to accurately synchronize clocks in different parts of the globe to establish geographical longitudes.[3] However, because the orbits of the planets in our Solar System are not confined to precisely the same plane, transits do not occur each time an inner planet overtakes the Earth in its orbit around the Sun. For Mercury, transits occur only about once a decade, and for Venus, they occur only a few times per century (Figure 13.1). When the Moon transits the face of our Sun, we can experience the beauty of a total solar eclipse as all the Sun's light is blocked for a few minutes. But when distant Mercury or Venus transit, they block only a few parts per hundred thousand of the Sun's light.

Hot Jupiters will produce transits for observers located within several degrees of their orbital planes. For a system like 51 Peg, an observer close to the orbital plane will see a transit every few days. During transit, such a giant planet blocks about one percent of the parent star's light and this can easily be detected with a modern electronic camera.

Consider a hypothetical planet in a 10-day orbit around a Sun-like star. For a random orientation of the orbital plane, there is a few percent chance that transits will be seen from our vantage point. In principle, one needs to monitor the brightness of fewer than a hundred stars with such planets to have a good chance of detecting at least one system in which the giant planet occasionally transits in front of its parent star. Since several percent of all stars appear to contain close-in hot Jupiters, the continuous monitoring

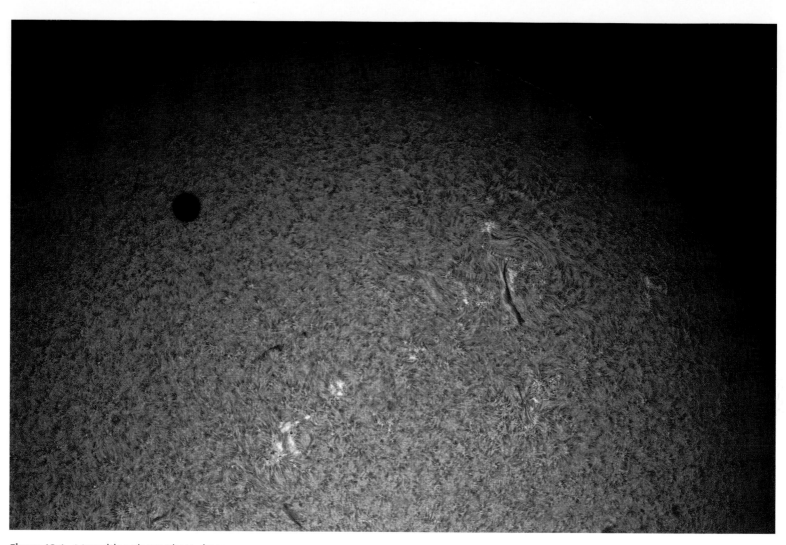

Figure 13.1. A transiting planet. The Hα image shows the 2004 transit of the planet Venus in front of the Sun. (Courtesy Stefan Seip).

of the brightness of ten thousand randomly chosen stars should suffice to discover many such systems.

Astronomers have by now detected several planets around distant stars using the transit method. In the late 1990s, several groups of astronomers established programs to look for transits from planets discovered using the radial velocity method. For such planets, both the orbital period and the time when a transit should occur are known. Transits are expected when the star is farthest from Earth (about one-quarter of an orbital period before or after the star exhibits its minimum radial velocity excursion from its average value). This knowledge enables a relatively efficient search for a transit signature. In 1999, Tim Brown and David Charbonneau found the first clear case of the long-sought transit signature in the light curve of a star known as HD 209458. They found a dip in the light of this star that repeated every 3.5 days: the orbital period of the suspected planet.[4]

Not all transits, however, are produced by planets. As discussed in earlier chapters, most stars live in binary star systems. If the orbital plane lies sufficiently close to our line of sight, these systems will undergo mutual eclipses. The naked-eye star known as Algol in the constellation Perseus is perhaps the best known such eclipsing binary. In many binary systems, the member stars have very different properties. For example, a Sun-like star may be orbited by a much fainter red dwarf, brown dwarf, white dwarf, or even a neutron star. These objects are so dim compared to typical main sequence stars that they do not contribute any easily detectable light. They can also be much smaller than the brighter star. Thus, when they transit, their signals may mimic that of a planet. As a result, searching for planets using the transit method can be tricky. Great care must be taken to weed out dwarf or collapsed companions. Fortunately, these objects are much more massive than planets, and they exert a large gravitational pull on their brighter companions. Stellar-mass companions can be distinguished from planets by the presence of large-amplitude reflex motion. Follow-up spectroscopic observations of suspected transiting planet candidates are essential.

Once a candidate planetary transit has been identified, intense monitoring can provide much information about the distant planetary system. For example, the interval between successive transits of a planet leads to a determination of the planet's orbital period. The duration of the transit and the shape of the transit light curve constrain the amount of light occulted by the planet, the size of the planet, and the inclination of the orbit. From the continued timing of transits, one can measure the shape of the orbit and perhaps detect the perturbations caused by the presence of other planets, or moons orbiting the planet. Since the planet is seen in silhouette against its parent star, careful observations of the star's spectrum during a planetary transit might reveal the presence of an atmosphere and lead to the determination of its properties through the tell-tale spectral signatures produced by atmospheric constituents.

Since transits can occur only for systems whose orbital planes lie near our line of sight, detection of a transit removes the major ambiguity of the radial velocity method: the orientation of the planetary orbital plane. If a planet transits, this orientation becomes determined: it must lie very close to our line of sight. The spectral type of the star, the radial velocity amplitude, and the orbital period can then be used to unambiguously determine the planet's mass. This method of analysis was applied to the first transiting planet, HD 209458, and revealed that it has a size of 1.27 times that of Jupiter (91 000 km) and a mass of 1.2×10^{30} grams (two-thirds that of Jupiter). These parameters imply that the transiting planet orbiting HD 209458 has an average density of only 0.38 grams per cubic centimeter. If there were an ocean large enough, this planet would float like a piece of cork!

Modern large-format CCD cameras can in principle monitor the light of millions of stars. Therefore, large-scale stellar-brightness monitoring campaigns have the potential of discovering thousands of planets. Even the typical amateur astronomer's telescope, equipped with a commercial CCD camera, and aided by perseverance, has the sensitivity to search for and detect hot Jupiters.

The transit method, when used from space, has the potential to discover even smaller, Earth-like planets. NASA is preparing to launch a satellite called Kepler to search for Earth-like planets using the transit method.

Gravity's lens

Einstein's theory of gravity provides us with another potential tool to use in the search for planets around other stars. Einstein predicted that light rays are bent by the pull of gravity. Instead of following a straight path, a light beam passing near a star or planet is deflected towards it. In 1919, Einstein's prediction was shown to be correct by Sir Arthur S. Eddington who obtained photographs of stars visible near the Sun during a total Solar eclipse. He compared their positions during the eclipse with positions determined from photographs obtained of the same star field about six months earlier when the Sun was located in the opposite portion of the sky. Eddington found that during the eclipse, the stars had shifted their apparent positions away from the center of the Sun by about 1.5 seconds of arc, about the apparent width of a thumb as seen from a distance of three kilometers. As the light from the distant stars passed near the Sun's limb, gravity pulled the rays slightly towards it. From the vantage point of an observer on Earth, the apparent positions of these stars were shifted *away* from the Sun; the amount of deflection was as Einstein predicted.

About 20 years ago, astronomers realized that the gravitational bending of light from a distant star by a foreground star may, on rare occasions, produce a large *increase* in the apparent brightness of the background object. Whenever a star moves within about an AU of our line of sight to a more distant star, the foreground star's gravity deflects the background star's light. Although we cannot see the tiny apparent shift in the position of the background star, it is easy to see it brighten.

If the two stars and the Earth were in perfect alignment, a very high resolution picture with a future space telescope would reveal that the image of the background star was distorted into a ring of light. This so called "Einstein ring" would have an apparent angular diameter of only a few *milli*-arcseconds: the apparent size of a thumb three *thousand* kilometers away. This distorted ring-like image is too small to be seen in any existing telescope. Because gravity bends light like a lens that creates an increase in apparent *brightness* of the background star, the term "gravitational lensing" has been coined.

Gravitational lensing has been seen in many astronomical situations. The large mass of a cluster of galaxies can distort and magnify the light of distant background galaxies, sometimes producing multiple and highly distorted images of these objects (see Figure 13.2). In a few very rare cases, the alignment between the foreground and background object is so perfect that we

Figure 13.2. A Hubble Space Telescope image showing gravitational lensing of a blue background galaxy by a foreground cluster of galaxies. Multiple images of the background galaxy are magnified and highly distorted into concentric arcs centered on the galaxy cluster. (NASA/STSCI/HST).

actually do see an Einstein ring many arc seconds in diameter. Because the star-on-star lensing produces much smaller Einstein rings than galaxies, they are usually referred to as gravitational *microlensing* events.

How does one look for a gravitational microlensing event? Since such events are very rare, one needs a dedicated telescope and a camera to take pictures of the same star field every clear night for many years. As stars move about the Galaxy, they typically take a few weeks to pass through the zone where gravitational lensing magnifies the light of the background star by about a factor of two. Even though such chance alignments are rare, they occur with sufficient frequency that about five microlensing events per year are seen for every one million stars being monitored. The odds of seeing a microlensing event for a particular star is quite a bit better than winning the lottery.

About a decade ago, Bohdan Paczynski and Andrew Gould realized that microlensing can also be used to search for planets. As already noted, for a microlensing event to take place, our line-of-sight must penetrate within about an AU of the foreground star. Fortunately, this is exactly the region where planets orbiting the foreground star are expected.

A planet orbiting a star that produces a microlensing event will generate an additional very short duration blip in the star's light curve *if* our line of sight also passes sufficiently close to the planet (about 0.1 AU for a giant planet, or 0.01 AU for a terrestrial planet). Paczynski and Gould demonstrated that for a planetary system like ours, there is a few percent chance that at least one planet would produce its own lensing signature on top of the microlensing signature of the parent star. While stellar microlensing events last for weeks, the additional brightening produced by a planet is expected to last only hours for a terrestrial planet, and a day or so for a gas giant like Jupiter. Therefore, to detect a planet, the light curve produced by the microlensing star must be sampled about once an hour. Thus far, only a handful of stellar microlensing events have been monitored with such vigilance, and so far only one planet may have been found by this method.

Despite the difficulties, gravitational lensing by planets is potentially one of the most powerful tools available in the quest for other worlds. Lensing events can be identified and monitored from either the ground by a network of modest aperture telescopes, or better, from space by a dedicated 1–2 meter diameter telescope. By expanding the scope of current searches to include hundreds of millions of background stars, we might identify hundreds of *stellar* microlensing events each year. By monitoring each of these events intensively, we have a means by which to detect planets down to the size of Mercury or even Pluto. Despite the difficulties, lensing provides one of the best hopes for detecting rocky terrestrial planets in orbit around other stars within the next decade. Microlensing may be used to determine the abundance of planets around random field stars in the Galaxy before the next generation of planet-hunting space telescopes become available.

The odd pulsar planets

The very first planetary-mass objects discovered outside our Solar System were actually found three years prior to the detection of the first hot Jupiters by Mayor and Queloz. In 1992, radio astronomers reported the discovery of three "planets" orbiting a *pulsar*. Pulsars are rapidly rotating neutron stars that emit periodic pulses of radio-wavelength energy towards us. They are the leftover remnants of massive stars that died in a supernova explosion, leaving behind an ultra-compact (roughly 10 km in size) sphere with a mass comparable to the Sun and consisting mostly of neutrons. Pulsars rotate with periods measured in fractions of a second. Like searchlight beacons, pulsars emit radio pulses at their spin rate. Accurate timing of the pulse arrival times provides a very sensitive method for detecting the reflex motion produced by any orbiting companion. The complex reflex motion of the pulsar known as PSR B1257+12 could only be understood if it were orbited by at least *three* objects having masses comparable to the terrestrial planets.

There is no question about the reality of these objects. But are they really planets? It is highly unlikely that any planetary objects could have survived in orbit around the progenitor star as it exploded. It is possible that the neutron star "stole" planets from another star in a close encounter. However, it is more likely that in the immediate aftermath of the explosion, and during the formation and early evolution of the neutron star that formed from the collapsing core of the massive star, some infalling material left over from the embers of the dying star formed a spinning accretion disk around the young pulsar. This disk then evolved and formed planet-mass objects in a manner similar to the birth of ordinary planetary systems. It appears that Nature has repeated herself around the collapsing core of a massive star that gave birth to a pulsar.

Extra-Solar gas giants and their host stars

A curious pattern has emerged from the investigation of the host stars to extra-Solar gas giants: they tend to have an exceptionally large abundance of elements heavier than helium. Thus, the probability of detecting gas giants increases with the abundance of heavy elements in the parent star. In this context, it is interesting to note that the Sun has a larger than average abundance of "metals." Astronomers call elements heavier than helium (e.g. carbon, nitrogen, or oxygen) a "metal" even though many are not metals in the chemical sense of the word.

There are several possible explanations for the correlation of extra-Solar planets with "metallicity." Since planets are known to contain an abundance of elements heavier than hydrogen and helium, planets may preferentially form from clouds with high "metallicity." Alternatively, it is possible that stars surrounded by planetary systems become polluted by infalling debris. Mature solar-mass stars do not mix matter that is dumped onto their surfaces into their interiors. Thus, if comets, asteroids, or rocky planets are accreted onto the surface of a star, the polluted layers would remain overabundant in metals compared to the rest of the star. Thus, high metallicity may be indirect evidence for contamination by infalling debris from a planetary system.

Planetary systems having a gas giant close to their parent stars probably preclude the existence of terrestrial planets in Earth-like orbits. The long-range gravitational tugs exerted by giants would expel any rocky planets that might have formed near them. In planetary systems dominated by gas giants in close-in orbits, such rocky planets might exist in more distant orbits. But from the point of view of their potential for life, far-out rocky planets are not good bets. They are much too cold to sustain liquid surface water, one of the essential ingredients of life as we know it.

Future telescopes in space and the quest for exo-planets

What are the prospects for finding Earth-like planets? NASA has ambitious plans for the next decades. In establishing its *Origins* program in the mid-1990s, Dan Goldin, NASA's Chief Administrator at the time, set forth the goal of building a series of ever more ambitious space telescopes designed to eventually produce pictures of nearby extra-Solar planetary systems. He challenged scientists to conceive and design the instruments needed to produce a picture of the surface features of an Earth-like planet orbiting another star within two or three decades. No telescope today is remotely up to this task. However, NASA has set in motion the first steps needed to meet this ambitious goal.

As discussed in Chapter 2, the *James Webb Space Telescope* is an ambitious 6 meter diameter instrument designed to work primarily in the infrared wavelength region.[5] Though the primary science goal for JWST is to study the birth and evolution of the first stars and galaxies as well as to investigate star and planet formation in the nearby cosmos, it may have the sensitivity and angular resolution to distinguish a Jupiter-like planet orbiting a nearby star.

Next is a series of innovative telescope concepts explicitly designed to search for Earth-like exo-planets orbiting nearby stars. NASA is actively developing a mission called *Terrestrial Planet Finder* (TPF).[6] One instrument for planet finding that has been proposed is an infrared wavelength interferometer which can combine the light of several small telescopes separated by tens or even hundreds of meters to *synthesize* the resolution of a much larger instrument. Modeling the performance of such an instrument in space indicates that it may be just possible to detect the feeble thermal-infrared radiation emitted by terrestrial and gas giant planets. Though such radiation is more than a million times dimmer than the infrared radiation emitted by the central star, interferometers consisting of several 1–2 meter diameter telescopes operated in concert from a space-borne platform may just have enough sensitivity to find Earth-like planets orbiting nearby stars. However, to build a TPF based on the principles of an infrared interferometer requires that we learn to fly a cluster of free-floating telescopes in formation and keep individual components positioned to a fraction of a micron when separated by hundreds of meters. This is a daunting technical challenge.

Another concept for planet finding is a single large-aperture (6–10 meter) telescope in space designed to block the light of the parent star, thus enabling astronomers to search for the dim reflected light of orbiting planets at visual wavelengths. But to detect a point of light so much fainter than its parent star requires an ultra-precise optical surface. The telescope has to be extremely free of stray light scattered from surface irregularities and mirror supports. Deviations of the mirror surface from its ideal shape that are larger than the size of an atom on scales of the order of 10 centimeters or so are unacceptable. Engineers think that they can build such a space-based telescope with the required precision using a "deformable mirror" to tune out the tiny errors inherent in even the most precise optics. By flying a 6–10 meter diameter telescope in space equipped with a deformable mirror and an occulting device, astronomers may be able to image and analyze the light of Earth-like planets out to a distance of about 30–100 light years from us. Such an optical TPF might be ready for launch by 2015 if funding is made available.

Which is easier: detecting exo-planets at infrared or visual wavelengths? The contrast between parent star and planet is less daunting at infrared wavelengths. Planets are about a million times dimmer than their parent stars at infrared wavelengths where they shine by emitted light; they are about a billion times fainter at visual wavelengths where they shine by reflected light. But the technology of infrared space interferometry is less mature than the technology of visual-wavelength telescopes. Though the contrast between star and planet may be a thousand times less favorable at visual wavelengths, the expected properties of Earth-like planets may make visual wavelength observations more desirable. The Earth's atmosphere is opaque at many infrared wavelengths, which is why we have to go to space to conduct sensitive infrared observations. While an infrared TPF would mostly probe only the stratosphere of such a planet, a visual wavelength TPF could see all the way down to its surface. Even though a distant planet may be merely a pale blue dot of light, analysis of its spectrum may reveal the tell-tale signs of oxygen, water vapor, liquid water, and continents on its surface. From the light curve of the reflected light, we may be able to determine the length of its day, and whether or not it has seasons. Should such an Earth-like planet harbor life similar to our own, it might be detected by the spectral signature of chlorophyl in the planet's reflected light.

Though it may be impossible to detect Earth-like planets with ground-based interferometers, such instruments may be able to detect the brighter near-infrared emission from gas giants in Jupiter-like orbits, especially if they are young. Young gas giants (with ages less than 100 million years) are both larger and warmer than mature planets. The twin 10 meter diameter Keck telescopes located on the summit of the Mauna Kea volcano in Hawaii are separated by about 100 meters. They were designed to be used together to form an interferometer. Also, the European Southern Observatory has built the *Very Large Telescope* consisting of four 8 meter diameter telescopes in northern Chile. The light they gather can be combined in a network of tunnels with the light from several smaller "outrigger" telescopes to form what is called the *Very Large Telescope Interferometer*. This giant instrument will eventually have the resolution equivalent to a telescope several hundred meters in diameter.

Is there any chance of meeting Dan Goldin's challenge? To obtain an image which can resolve 100 kilometer features on a planet 30 light years from the Sun is equivalent to seeing a thumb from half way to the Sun, or reading a newspaper headline at a distance of 50 million miles. To attain such incredible resolution at visual wavelengths requires an instrument having a baseline of about 2000 kilometers! Alternatively one could build a space interferometer that works at wavelengths twenty times longer in the infrared. Near a wavelength of 10 microns, planets produce their own heat radiation. Such an instrument requires a 40 000 kilometer baseline to provide a resolution of 100 kilometers on a distant planet. Perhaps in a future century, a squadron of precision telescopes might be built. How much would such an instrument cost? By one estimate, the "Goldin Dream" would cost at least 500 *billion* dollars, comparable to the budgets of many nations.

We live in an exciting time for planetary astronomy. Astronomers have found indirect evidence for planets around more than 100 stars. As we build more powerful instruments, we may find that the percentage of stars with planets is even larger. The discovery rate already implies that at least 5–10 percent of stars in the Galaxy are likely to have planetary systems. There could be several billion planetary systems in our Galaxy!

Part III The cosmic context

Cosmic cycles

14

Stellar death plays a fundamental role in the birth of new generations of stars. Stars produce most of the elements heavier than hydrogen in thermonuclear fusion reactions in their cores. During the course of stellar evolution and death, much of this material is expelled back into interstellar space. As a result, the interstellar medium becomes progressively enriched with an ever increasing abundance of various elements. Stars also release vast amounts of energy in the form of light, winds, and explosions. The physical state of the interstellar medium is to a large measure regulated by the processing of this energy. In this chapter, we will first review how long-lived low-mass stars and short-lived massive stars evolve and die. Then we discuss how stars re-cycle much of their matter, and how they may induce the formation of new molecular clouds.

Light, winds, explosions

As we saw in previous chapters, massive stars create HII regions that efficiently destroy clouds and profoundly impact the formation of lower mass stars and their planetary systems. Massive stars can also trigger stellar birth. The exploding remnants of clusters of massive stars in OB associations create bubbles which can sweep up millions of solar masses of relatively low density interstellar matter into giant shells and rings hundreds or even thousands of light years in diameter. Over tens of millions of years, these shells decelerate, fragment, and sometimes collapse into new clouds. Calculations show that near the Sun, the first fragments to collapse under the force of gravity have masses comparable to giant molecular clouds. There exists a grand 50–100 million year cycle of star formation, cloud destruction, massive star evolution, cataclysmic death in supernova explosion, giant bubble expansion, gas-shell gestation, cloud condensation, followed by more star formation. Such is the great "galactic ecology" of stellar birth, life, and death.

Evolution and death of low- to intermediate-mass stars

Hydrogen in a stellar core is depleted by the gradual conversion into helium.[1] The accumulated helium is compressed and heated by gravity until a new thermonuclear fusion reaction, the *triple alpha* process, ignites helium and burns it to carbon.[2] As the star enters this phase, the extra energy generation causes it to swell. The star becomes brighter (it has more surface area to radiate from) and its surface layers become somewhat cooler (it becomes redder). A star in this phase is said to be evolving towards the *red giant* phase.

All stars below about eight times the mass of the Sun eventually evolve into red giants. They develop strong, but relatively low-speed (10–50 km/s) winds which can remove much of the red giant's outer envelope in a few million years. Some red giants shed their outer layers into slightly irregular envelopes. Others expel most of their material into an equatorial torus. As they exhaust their thermonuclear fuel and loose their surface layers, their dense cores become visible as hot white dwarf stars. White dwarfs are no larger than terrestrial planets, but contain a large fraction of the star's initial mass.

Like massive OB stars, young white dwarfs emit most of their energy in the ultraviolet portion of the spectrum. They also power fast stellar winds which plow into the slowly expanding debris ejected during the earlier red giant phase. The interiors of these envelopes become ionized by radiation and shocked by fast winds (Figure 14.1). These HII regions, created by white dwarf stars forming from the exhausted cores of dying low-mass stars, are the planetary nebulae already mentioned in Chapter 8. Sometimes the radiation and winds of the white dwarf can attain a bipolar pattern. The resulting collimated outflows are reminiscent of the flows produced by forming stars. Some planetary nebulae even have highly collimated jets whose internal shocks resemble the Herbig-Haro objects discussed in Chapter 6. While planetary nebulae mark the births of white dwarf stars, they also represent the relatively gentle deaths of low- to intermediate-mass stars (Figures 14.2 and 14.3).

The Sun will become a red giant in another 5 billion years. It will swell to nearly the orbit of Mars, in the process swallowing the Earth, and become about ten thousand times as luminous as it is today. Much of its outer layers will be ejected, recycling into the interstellar medium some of the products of its thermonuclear fires. Its core will eventually collapse to form a white dwarf. UV radiation and winds will ionize and shock the interior of the red giant envelope, forming a planetary nebula which will glow for about a hundred thousand years. The Sun will leave behind a white dwarf remnant that will fade over billions of years as its residual heat is radiated away into space.

No star with a birth mass less than about 0.8 times the mass of the Sun has yet died of natural causes in this Universe.[3] Because such stars live longer than 14 billion years, none have yet exhausted their supply of thermonuclear fuel.

White dwarf stars have a maximum mass of about 1.4 solar masses. For larger masses, the electron pressure responsible for stabilizing these

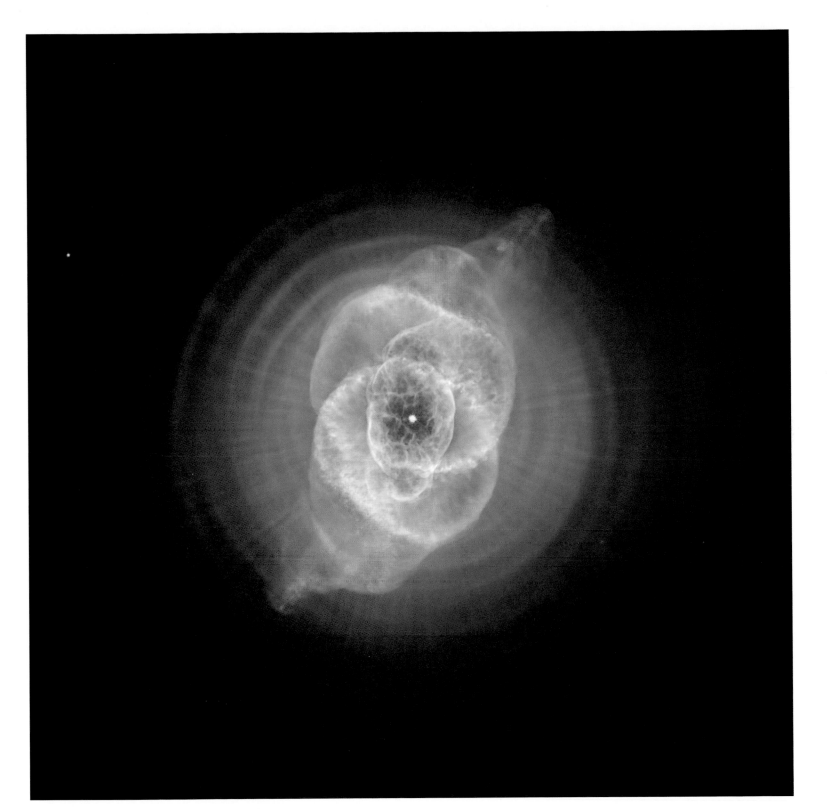

Figure 14.1. The planetary nebula NGC 6543, also
known as the "Cat's Eye Nebula," shows intricate
structure with concentric shells, jets, and shocks.
The nebula is about 1000 years old and is a signpost
of a dying star. (NASA/STScI).

Figure 14.2. The Egg Nebula in Cygnus is a Sun-like star evolving towards the planetary nebula stage. (NASA/STScI).

collapsed stellar cores is overwhelmed by the force of gravity. As discussed in previous chapters, many stars are born in binary or multiple star systems. The more massive member of low-mass binary star systems will be the first to become a red giant, form a planetary nebula, and produce a white dwarf. The brightest star in the sky, Sirius, has such a white dwarf companion. Eventually, at the end of its life, the second star goes through the same evolutionary stages. As the secondary star in a close binary containing a white dwarf swells into a red giant, it can transfer substantial amounts of its mass onto the white dwarf companion.

Mass-transfer binaries are responsible for several varieties of variable star in the sky, including cataclysmic variables that occasionally flare in brightness, *nova* outbursts during which the star increases its brightness by factors of thousands for weeks at a time, and the type Ia *supernova* explosions in which a white dwarf detonates in a cataclysmic thermonuclear explosion. As accretion from a companion pushes the mass of the white dwarf above 1.4 solar masses, gravity overwhelms the electron pressure supporting the star and it collapses. Increasing density and heat ignites violent thermonuclear burning of carbon, helium, and residual hydrogen that rips the white dwarf apart in an instant.

Evolution and death of massive Stars

Massive stars are short lived, but very energetic, as we have seen in previous chapters. Like their lower mass siblings, high-mass stars also produce jets, winds, and eruptions during their birth. But massive stars form and evolve much faster. In youth, massive stars produce outflows that are far more powerful than those of their lower mass counterparts. As these stars grow to more than eight times the mass of our Sun, they start to produce intense

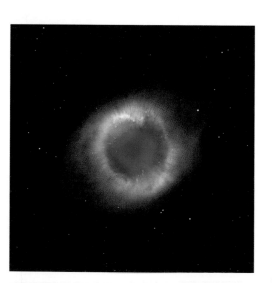

Figure 14.3. The Helix Nebula is one of the nearest and largest planetary nebulae, covering more than half the Moon's size on the sky. (NASA/ESA/STScI/NOAO).

Figure 14.4. The Bubble Nebula is a shell of gas and dust carved out by the stellar wind of a massive star. (D. Williams, NOAO/AURA/NSF).

ultraviolet light which floods their environment. Ionizing radiation produces compact HII regions that carve large cavities in the surrounding cloud. By breaking the bonds that tie atoms together to form molecules, the massive young stars transform much of the surrounding molecular cloud from which they were formed into expanding bubbles of hot plasma. These luminous and massive stars destroy their parent clouds. The birth, evolution, and impacts of expanding ionized nebulae were discussed in Chapters 9 and 12.

While the protostellar outflows from low-mass stars abate as the stars evolve towards the main sequence, the most massive stars continue to blow powerful *stellar winds* with speeds of thousands of kilometers per second into the ionized nebulae created by their UV radiation. Although low-mass stars also blow winds, they are feeble in comparison to those produced by the massive stars. For example, the most luminous member of the Trapezium cluster in Orion blows a wind a million times stronger than our Sun's comparative breeze.[4] Like escaping steam rising from a boiling pot that gradually loses its contents, stellar winds can remove most of a high-mass star's material over its lifetime. These stellar winds slam into the surrounding ionized nebulae and carve out *stellar wind bubbles.*

As a stellar wind encounters denser gas or other obstacles such as small clouds, powerful shock waves generate a tenuous plasma with temperatures of *millions* of degrees, almost as hot as the interiors of normal stars. Thus, the shocked gas in wind bubbles emits X-ray radiation. Such hard radiation is absorbed by our atmosphere and can only be observed with satellites like the *Chandra* X-ray Observatory. When the medium into which a stellar wind bubble propagates is sufficiently dense, the advancing shock wave can sometimes be seen in optical images as gossamer filaments of glowing hydrogen and doubly ionized oxygen. Figures 14.4 and 14.5 show examples of stellar wind bubbles at visual wavelengths.

Sun-like stars do not have enough mass to compress their stellar cores sufficiently to ignite the thermonuclear reactions leading to elements heavier than carbon. However, the more massive stars do. Like their low-mass kin, massive stars live their main sequence lives by burning hydrogen into helium in a thermonuclear fusion reaction. But massive stars do it much faster. As a high-mass star evolves towards the *red supergiant* phase (the equivalent to the red giant phase for low-mass stars), a core dominated by helium develops underneath the outer layer of hydrogen. As it compresses and heats, the helium core reacts to form carbon, followed by additional thermonuclear reactions that consume carbon. Carbon and residual helium nuclei in the core react to produce oxygen, and oxygen in turn reacts with helium to produce neon, and neon reacts with helium to form magnesium, and magnesium burns to silicon, and so on up to iron. In successive steps, the chemical elements lighter than iron are formed. The massive stellar core comes to resemble an onion consisting of nested shells. Each shell is dominated by a heavy element synthesized from the thermonuclear burning of the adjacent outer shell.

Figure 14.5. The Crescent Nebula, NGC6888, is a wind bubble carved by mass-loss from a massive star located in the nebula's center. (T.A. Rector, NOAO/AURA/NSF).

Like stars of lower mass, red supergiants also lose mass. But they do so at a feverish pace. The most massive stars can completely lose their outer hydrogen-rich envelopes and expose their helium rich cores. Some even bare their carbon- and nitrogen-rich interiors. These supergiants can attain 10 million times the luminosity of the Sun and become the most luminous stars in a galaxy. As these luminaries evolve, they alternate between cool red supergiant and hot blue supergiant phases.[5] Intense radiation fields exert a powerful outward force on their outer layers. Occasionally, radiation pressure exceeds the pull of gravity and the star's outer layers are violently ejected. The remnants of such eruptions produce spectacular bubbles of expanding plasma and giant ring nebulae.

The star Eta Carinae, one of the most massive stars in the Milky Way, underwent a great eruption in the 1840s. For about a decade, it shone as the second brightest star in the sky, only to fade from view as dust condensed in the several solar masses of ejecta that was launched into space with speeds of over 600 km/s (Figures 14.6 and 14.7). Today, this expanding envelope is, at some infrared wavelengths, the brightest source in the sky after the Sun. The thermonuclear evolution of aging stars more massive than about eight

Figure 14.6. The Carina Nebula surrounds a group of the most massive stars in our Galaxy, whose strong ultraviolet radiation lights up the surrounding gas. (N. Smith & J. Bally).

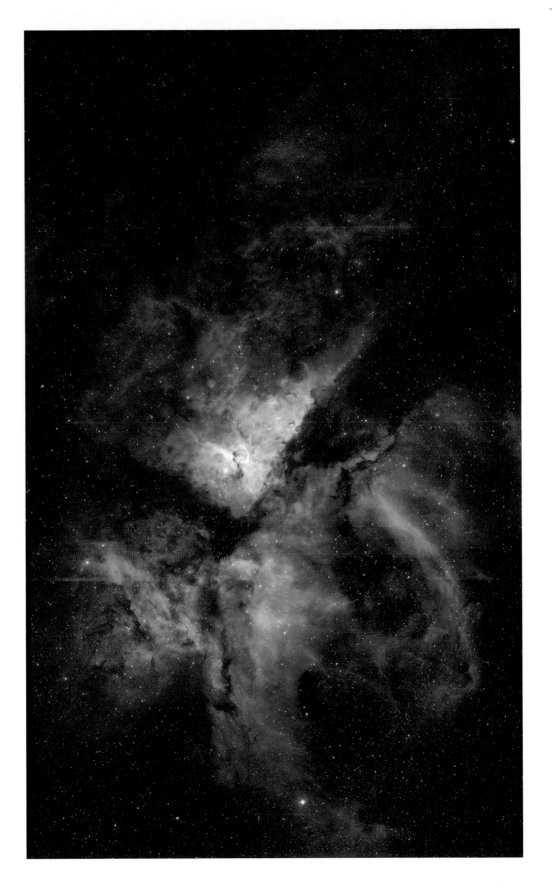

Figure 14.7. At the center of the Carina Nebula lies the star Eta Carinae, which is surrounded by a cloud of expanding gas from an eruption in 1841, when the star was the second brightest in the sky. (N. Smith, NASA/STScI).

solar masses continues until the element iron is synthesized in the very core of the star.

The accumulation of an iron core marks the end of the thermonuclear life of a massive star. Iron is at the bottom of the nuclear binding energy curve.[6] Thermonuclear fusion reactions that build heavier elements from lighter ones can generate energy only until the formation of iron. The formation of more massive elements *consumes* energy. Thus, as an iron core accumulates in the core of a massive star, the star faces an energy crisis. When the remaining nuclear fuel from its outer layers is consumed, the star runs out of thermonuclear energy. As the star's surface radiates away internal heat, its iron rich core shrinks under the overlying weight. At the end of the massive star's life, the core collapses and the star dies in a cataclysmic supernova explosion.[7]

During the collapse, increasing pressure and density squeezes electrons into the protons contained in iron (and other) nuclei. The result is an incredibly dense sea of neutrons packed together so tightly that a mere cubic centimeter can weigh as much as a large ship.

The collapse of the iron core of a massive star takes only a few seconds. And in those moments, the formation of a neutron star releases more gravitational potential energy than the total amount of energy released by the Sun over its entire 10 billion year life.

Like for white dwarfs, there is an upper mass limit for neutron stars. Neutrons stars with more than about five times the mass of the Sun are not stable. Even neutrons cannot support the tremendous weight of such a star. If the collapsing stellar core contains more than about five solar masses, it may form a black hole.

As stars evolve toward their end-states, they lose most of their mass. Stars with initial masses less than about eight times the Sun eventually die as white dwarfs with masses less than about 1.4 solar masses. Stars with initial masses between 8 and 40–50 solar masses explode as supernovae that leave behind neutron star remnants with masses below about five solar masses. Stars born with even larger masses are thought to form black holes during their terminal supernova explosions.

The tremendous release of gravitational potential energy resulting from the formation of a neutron star or a black hole ejects the outer layers of the star. The collapse of the stellar core allows the outer layers of the star to fall in. However, as this material encounters the ultra-dense iron core or its neutron star remnant, it "bounces." Implosion becomes explosion accelerated by the energy released by the forming neutron star. Compression and heating, first produced by infall motions and later by the shock waves formed as expanding post-bounce material runs into the dying star's outer layers, drives the energy-consuming thermonuclear reactions which forge the elements beyond iron. As these shock waves break through the stellar surface, the dying star's outer layers can be accelerated to nearly 10 percent of the speed of light. A few hours later, the exploding debris reaches the brilliance of a 100 billion Suns! The dying star may outshine its host galaxy with its billions of stars for several weeks.

Supernovae have profound impacts on their surroundings. The exploding star's debris blasts through the layers ejected during its supergiant phase, its stellar wind bubble, and the remnants of its ionized nebula. The expanding debris sweeps up surrounding gases and forms a supernova remnant which can grow to sizes of hundreds of light years. The supernova explosion provides one of the most important mechanisms by which the products of thermonuclear fusion are fed back to enrich the interstellar medium of a galaxy with elements heavier than helium.

Supernova remnants, superbubbles, massive shells, and giant rings

The first stellar explosion in an OB association will blast out a cavity and sweep up a dense shell of gas and dust. But as the supernova remnant sweeps up its surroundings, it decelerates. By the time they reach light-year

Figure 14.8. The Crab Nebula is the remnant of a supernova that exploded in 1054 when it became almost as brillant as the Moon. (ESO/VLT).

dimensions, most observed supernova remnants have decelerated to speeds of hundreds of kilometers per second. Their shock waves become visible throughout the spectrum as they accelerate electrons and low-energy cosmic rays and excite atoms and ions. The charged particles gyrate in the feeble interstellar magnetic field and the field carried by the expanding supernova debris. Energetic electrons emit radio waves; supernova remnants are prodigious sources of such *synchrotron radiation*. The swept-up interstellar gas and supernova ejecta emit in the emission lines typical of shock-waves. Though similar in nature to the shocks associated with Herbig-Haro objects, which were discussed in Chapter 6, supernova shocks tend to be faster and much larger.[8]

The best known supernova remnant in our Galaxy is the famous Crab Nebula in Taurus. At the location of the Crab Nebula, a supernova exploded in the year 1054 and was recorded by Chinese astronomers. A glowing network of hydrogen filaments and knots today traces the expanding debris of the shattered star (Figure 14.8). This debris is embedded in a veil of synchrotron radiation at wavelengths extending from the X-ray to the radio parts of the spectrum. In the center, a *pulsar* flashes its searchlight beam of

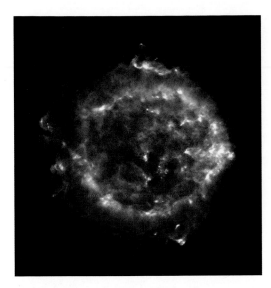

Figure 14.9. A radio image showing the synchrotron emission from the Cassiopeia A supernova remnant that exploded 300 years ago. (NRAO/AUI).

radio waves in our direction 33 times a second. An even younger supernova remnant is located in the constellation Cassiopeia and results from an eruption in the seventeenth century. The event was not seen at the time, since the area has large dust obscuration. The hot plasma in the supernova remnant emits copious amounts of X-ray emission. An image obtained at radio wavelengths is shown in Figure 14.9. Radio astronomers have found several hundred supernova remnants throughout our Galaxy. Most are shells hundreds of light years in diameter. The Veil Nebula in Cygnus, a wreath-like network of glowing hydrogen, ionized oxygen, and sulfur filaments scattered over a several-degree diameter patch of sky, is a roughly 10 000-year-old remnant of an ancient massive star explosion (Figure 14.10).

From their numbers, expansion speeds, and sizes, astronomers estimate that the birth-rate of supernova remnants is about one every 50 years in our Galaxy. This is a lower bound on the rate of supernova explosions (the rate at which massive stars in our Milky Way are born and die) because not all supernovae produce detectable remnants. Some may be smothered inside dense molecular clouds. Others may explode in such tenuous environments that they fail to produce visible shocks or radio shells.

As supernova remnants sweep up interstellar material, they slow down, and their radio, optical, and X-ray emission fades. Although they typically fade within about 100 000 years, their swept-up shells of dust and hydrogen act as signposts of explosions for millions of years (Figure 14.11). The northern part of Orion contains a ring of dust and atomic hydrogen emission over 6 degrees in diameter. Centered on the star Lambda Orionis, this ring marks the demise of a massive star in northern Orion about a million years ago (see Figures 9.4, 9.5 and 9.6).

About 10–30 percent of massive stars are low-velocity runaways with speeds of tens of kilometers per second that were probably ejected by dynamical interactions in clusters or multiple star systems. Another 5–10 percent are fast runaways produced by the breakup of close binaries in which the more massive member exploded. Moving with speeds of hundreds of kilometers per second, such runaway stars can move thousands of light years from their birthsites before exploding. Therefore, while the majority of massive stars explode close to home, fast runaway stars seed the interstellar medium with a background of random supernova remnants. The demise of white dwarfs in Type I supernova explosions is responsible for an additional background of randomly located explosions.

The most common type of massive star (a roughly eight solar mass object[9]) lives about 30–40 million years prior to exploding. The majority of stars drift away from their birth sites with speeds of a few kilometers per second. At a typical speed of 3 km/s, a star moves about 300 light years from its birthsite by the time it explodes. The shorter-lived more massive stars tend to stay closer to home.

The first supernova in an OB association is most likely to mark the death of its most massive star. Subsequent explosions of lower mass stars inject

Figure 14.10. The Veil Nebula is part of a supernova remnant from a star that exploded about 10 000 years ago in the constellation Cygnus. (T.A. Rector, NOAO/AURA/WIYN).

additional stellar debris into the void left behind by the first. As debris catches up to the older shell, its impact can rejuvenate the older remnant, expanding it further. Thus, the massive stars that die in OB associations collectively inflate giant bubbles energized by many individual supernova explosions. In the larger star forming regions such as 30 Doradus, these *superbubbles* are driven by the combined effects of hundreds of massive stars. Inflated by ionizing radiation, stellar winds, and supernovae, they can grow to thousands of light years in diameter. Since shells older than about

Figure 14.11. A combined radio and far-infrared map of a star-forming region in Cygnus reveals cocoons of heated gas around newborn massive stars and expanding supernova remnents, which contribute to the recycling of the interstellar medium in this region. (J. English & R. Taylor, DRAO/IRAS).

30 or 40 million years are no longer energized by additional explosions, they decelerate and their interiors cool.

Like individual supernova remnants, superbubbles sweep up low density matter between the stars, much as a snowplow sweeps a street after a snow storm. The resulting giant shells, called *supershells*, can accumulate millions of solar masses of interstellar matter. The disks of the Milky Way and other galaxies are peppered with hundreds of such superbubbles rimmed by supershells.

The nearest OB association is found in the constellations of Scorpius and Centarus in the southern sky. Located at a distance of about 450 light years, dozens of massive stars visible to the naked eye are sprinkled over a 50 degree portion of the sky. There are thousands of much fainter lower mass young stars accompanying the massive luminaries. Like the Orion OB association discussed in Chapter 9, the most massive stars in the Sco-Cen group died over the last 10 million years. Their supernova explosions inflated a superbubble visible from our vantage point as a giant ring of atomic hydrogen more than 100 degrees in diameter. The Sco-Cen OB association is so close that the near wall of this bubble has already flowed around the Solar System. The debris in the superbubble interior blows past the Sun with a speed of 20 km/s, forming an interstellar wind of hot, very low density plasma coming towards us from the general direction of the Sco-Cen stars.

Another group of O and B stars is currently forming a relatively small superbubble in the constellation of Perseus. Located at a distance of about 900 light years, the Perseus OB2 association has spawned a 20 degree diameter ring of atomic hydrogen in this part of the sky. The Perseus molecular cloud mentioned in earlier chapters marks the location of ongoing star formation in this association.

Perhaps the most studied superbubble is the one created by the Orion OB association located 1500 light years from us and already discussed in Chapter 9. Star formation in the Orion region over the last 10 to 15 million years spawned dozens of massive stars, many of which have already exploded as supernovae. The inner rim of the Orion supershell is lit up by the remaining massive stars that are still shining in Orion (Figure 9.8). The brightest part of this feature is traced by the glowing red hydrogen crescent which wraps around the eastern rim of the Orion constellation and is known as Barnard's Loop. Like hot gas rising in a chimney, the supernova-heated plasma in Orion is expanding away from the plane of the Milky Way towards the constellation of Eridanus, located some 40 degrees west of Orion. A faint hydrogen feature called the Eridanus Loop marks this portion of Orion's supershell and is over 900 light years from Orion's active sites of ongoing star formation. The Orion superbubble is blowing towards us: its near wall is less than 500 light years from us. Cool dust and 21-centimeter radio emission from atomic hydrogen mark the outer walls of Orion's supershell. Infrared and radio telescopes show that Orion's superbubble is over a thousand light years in extent and is billowing mostly away from the dense mid-plane of our Galaxy.

Superbubbles can grow larger than the thickness of the gas layer in the Galactic disk. As their shocks race into the lower density medium above and below the Galactic plane, their hot interiors distort and, like volcanos, the largest bubbles erupt from the Galactic disk. They vent hot, shock-heated plasma from their interiors into the near vacuum above and below the Galactic disk. But like a geyser, the gravity of the Galactic disk eventually pulls much of this material back. Thus, about 30 to 60 million years after the formation of a superbubble, debris lofted out of the disk rains down on the mid-plane.

While a superbubble can vent above and below the Galactic disk, its sideways expansion is blocked by the denser gas in the Galaxy's plane. In this direction, the superbubble sweeps out a dense ring of slowly expanding material containing millions of solar masses of gas. Under the force of their own gravity, some of these dense ring-structures may fragment into new giant molecular clouds.

Formation of molecular clouds: a new beginning

The emerging picture of star formation will not be complete until the formation, evolution, and destruction of molecular clouds is understood. In previous chapters, we saw how jets and outflows produced by low-mass young stellar objects churn molecular clouds. But the greatest damage is done by massive stars whose UV radiation, stellar winds, and terminal supernova explosions can dissociate, ionize, and disrupt molecular clouds. Thus, the death and destruction of molecular clouds is a direct consequence of star formation. However, the formation of clouds and their evolution are still poorly understood and are active areas of research. In this section, we will explore some of the ideas of cloud formation and evolution.

How are molecular clouds created? This question has vexed astronomers for decades. Some have argued that the low density interstellar medium slowly condenses under the force of its own gravity to form new molecular clouds. Computer models show that the cooling of interstellar gas can lead to the formation of large evacuated voids surrounded by a filamentary network of cooler and denser gas. The densest portions of such filaments may be compressed by the hotter surrounding medium until self-gravity takes over, dust shields the cloud interior from UV radiation, and molecules form. As we have seen, however, the superbubbles created by OB associations born from previous generations of clouds sweep up much of the interstellar gas into dense shells and rings in the Galactic plane. Thus, the supershells produced by OB associations interfere with the quiescent condensation of clouds from interstellar space.

Another suggestion is that clouds form as the spiral arms of our Galaxy (to be discussed in Chapter 15) sweep up and compress interstellar gas. This model finds strong support in the spectacular arms of spiral galaxies such as Messier 51 and Messier 81. In these galaxies, the young blue stars, ionized nebulae, and associated dark molecular clouds are tightly concentrated into a clear spiral pattern. While this mechanism may explain the formation of some clouds, it does not easily explain the pattern of clouds and star formation around the Sun.

Yet another idea is that the infall of large atomic hydrogen clouds from outside our Galaxy might trigger the formation of some molecular clouds. The Milky Way Galaxy is surrounded by a network of tenuous atomic hydrogen clouds far above and below the Galactic plane. Many of these clouds approach the Galactic plane with speeds of more than a hundred kilometers per second. One source of high-velocity clouds is the tail of tidal debris torn out of our closest extragalactic neighbors: the Large and Small Magellanic Clouds visible from the southern hemisphere. These small, irregular galaxies orbit our Milky Way as satellites. Other high-velocity clouds may trace debris that was initially launched by superbubbles. But the origin of most high-latitude clouds is not yet understood. These hydrogen clouds collide on occasion with the Galactic plane. Such collisions are, however, too infrequent to trigger the birth of most giant molecular clouds.

Finally, the history of star formation in the Sun's vicinity over the past 50 million years suggests that the molecular clouds near us formed from the fragmenting walls of an ancient supershell. Gravity can cause the massive rings of gas swept up by old superbubbles to break up into individual clouds. Models of this process show that the masses of such ring fragments are comparable to the observed giant molecular clouds.

It is possible that all of these different mechanisms contribute in some way to the formation of molecular clouds. But how do we disentangle the roles played by these various processes? One approach is to look for clues in the fossil record of recent star formation surrounding the Sun.

The Gould's Belt: a fossil record of star formation

Well over a century ago, it was noted that the distribution of many bright blue stars (which we know today are relatively massive, hot, and therefore short lived compared with the Sun) have a peculiar tilted distribution with respect to the plane of the Milky Way. These bright blue stars follow a band all around the sky that defines a plane tilted by about 15 or 20 degrees with respect to the Milky Way. This system of blue young stars, called the "Gould's Belt",[10] is dominated by Orion's young stars in the winter sky and by the nearest OB association, the Sco-Cen OB association, in the summer constellations of Scorpius and Centarus. While the more distant Orion OB association is centered about 15 degrees below the Galactic plane, Sco-Cen is located well above the plane.

In the 1950s, the first high-quality radio-wavelength maps of the distribution of the 21 centimeter spectral line of atomic hydrogen[11] were obtained with radio telescopes. These observations revealed an expanding ring of atomic hydrogen roughly aligned with Gould's Belt of blue stars. This curious feature in the distribution of nearby cold gas came to be known as Lindblad's ring. About 35 years later, during the 1980s, shorter wavelength radio studies of carbon monoxide demonstrated that most nearby molecular clouds are also associated with the Lindblad hydrogen ring and Gould's Belt of young stars. The Sun is located off-center but inside this ring. The ring is elliptical with a major axis dimension of about 1800 light years and a minor axis dimension of about 900 light years and is expanding with an average speed of about 3 km/s from a location in the constellation of Perseus about 500 light years from us.

Stars somewhat older than those in Orion, Sco-Cen, and Perseus are found near the evacuated center of the Lindblad ring and the Gould Belt. In the 1940s, Adrian Blaauw found a group of stars centered about 500 light years from us that share a common motion through space and which have roughly the same ages. Because they are located towards the constellations of Cassiopeia, Perseus, and Taurus, he named these stars the "Cas-Tau Group." A bound open cluster of stars, the Alpha Persei cluster, lies at the center of this group. The most massive star today in this group has a mass less than seven times that of the Sun. But the large numbers of low-mass stars indicate that, originally, there must have been dozens of massive stars in this association, all of which have died. Studies of the remaining stars reveal that the group has an age of 40 to 90 million years. Because all stars sufficiently massive to explode as supernovae have already died, Blaauw dubbed the Cas-Tau group a *fossil* OB association.

Above and below the Alpha Persei cluster, there is an unusually low amount of interstellar gas and dust. This may be evidence that the superbubble created by Blaauw's stellar group blew out of the Galactic plane. This hole in our Galactic sky may be the remnant of the chimney which vented the energy released by dozens of high-mass stars born in this group.

From the ashes of the old: new clouds, new stars

An intriguing model of star and cloud formation has emerged from the study of the distribution of stars younger than about 90 million years and from the location and motion of the interstellar medium in the solar vicinity. In our backyard, the fragmentation of the supershell powered by the Cas-Tau fossil OB association may have formed a ring of giant molecular clouds that produced the Gould's Belt of stars. In a grand 50 to 100 million year cosmic cycle, these clouds are now forming a new generation of OB associations in Orion, Perseus, and Sco-Cen. In the following, we examine this scenario in more detail.

It all began with an unusually large molecular cloud that spawned a big OB association about 50 to 100 million years ago, the Cas-Tau group. Perhaps the birth of this cloud and its stars was triggered by the passage of a spiral arm of the Galaxy. In such regions, cloud and star formation may be enhanced by higher pressure and the increased concentration of clouds.

Within 10 or 20 million years, the parent cloud was disrupted by expanding blisters of ionized plasma created by its litter of massive stars. As the most massive stars exploded, their remnants merged to form a growing superbubble. The dispersed remnants of the parent cloud and surrounding low density interstellar gas were swept up into an expanding shell. Where the superbubble ran into the Galactic plane, its shell decelerated. But as it expanded into the void above and below the plane, there was less matter to slow it down and the superbubble created fountains of plasma which blew out on either side of the Galaxy. Astronomers have found a region right above the Cas-Tau group which has so little obscuration that it provides a very clear view of distant galaxies. This so-called *Baade's Hole*[12] is a result of the Cas-Tau superbubble blowing out of the Galactic plane.

About 30 million years after the end of star formation in the Cas-Tau group, the last supernova exploded. From then on, the superbubble decelerated as it ran into ever more interstellar matter. Plasma ejected above and below the plane by the *galactic fountain* cooled and condensed into clouds of atomic hydrogen. The gravitational pull of our Galaxy's disk has stopped and reversed the motion of some of these hydrogen clouds. They are now falling back towards the plane.

About 20 million years ago, the massive ring swept out of the galactic plane by the superbubble contained several million solar masses of gas. The ring became unstable to its own self-gravity and started to fragment into a ring of new molecular clouds. Calculations show that in the solar vicinity, the first self-gravitating fragments of a super-ring have masses around 100 000 times that of the Sun, comparable to typical giant molecular clouds.[13] At the Sun's distance from the Galactic center, gravitational instability of super-rings sets in just as clouds launched above and below the Galactic plane by the Galactic fountain begin to rain back down. This infalling debris reinforces the compression of clouds by their self-gravity. As dust grains shield cloud interiors from starlight, molecules form.

New star formation has ignited in several of these clouds and the resulting OB associations delineate the Gould's Belt of stars. Perhaps some of these new stellar groups will eventually sweep up massive super-rings which in turn will spawn new clouds, and future OB associations.

Thus, there is a 50 to 100 million year cycle starting with the formation of molecular clouds and OB associations, the destruction of these clouds by massive stars, the formation of superbubbles and massive super-rings, and ending with the onset of gravitational instability in the rings leading to the birth of a new generation of clouds. Molecular clouds typically convert only a few percent of their mass into stars prior to disruption. Any particular atom may loop around this cycle dozens of times over the course of several billion years before being incorporated into a star. If it finds its way into a massive star, the atom may be recycled into interstellar space within 30 million years, possibly converted into a heavy element. But sooner or later it will be incorporated into a long-lived low-mass star, and thus be removed from the cycle of star formation.

Without the infall of fresh gas from outside the Galaxy, the great cosmic cycle of cloud and star formation, stellar death, and cloud re-formation will eventually deplete the interstellar medium of our Galaxy. As a result, billions of years from now, star formation in the Milky Way will cease. In the distant future, as the more massive, blue, and brilliant stars die, the dim, red, and long-lived low-mass stars will come to dominate our fading Milky Way.

15 Star formation in galaxies

As we explore the Universe with our telescopes, it seems that we always find objects and phenomena which are larger, better, bigger, or more extreme. So it is with star formation. We have discovered that most stars form as multiple systems in very dense but transient clusters which are gravitationally unbound. These young star clusters disperse into the disk of our Galaxy as either T or OB associations soon after birth. However, occasionally star formation is efficient enough to leave behind a system of stars whose self-gravity prevents dispersal of the cluster members into the field. The result is usually an open cluster. When we look far and wide in the Universe, we see titanic bursts of star formation which dwarf all star-forming regions currently seen in the Milky Way. These "super-star-cluster" forming events may produce globular clusters and may populate the rich star fields of surrounding galactic nuclei. In this chapter, we examine such extreme cases of star formation in other galaxies.

Giant islands in the sky: the galaxies

For two centuries, astronomers have known about the *spiral nebulae*, faint and diffuse whirlpools of light sprinkled randomly across the sky. Yet, their nature remained a mystery until the 1920s. The interpretation of these objects sparked heated debates among scientists during the early part of the twentieth century. Today, we know these objects as *galaxies*, giant systems of stars and gas similar to our own Milky Way. Deep surveys of the sky indicate that there may be about a thousand billion galaxies. There are as many galaxies in the Universe as there are stars in large galaxies.

Galaxies are made of billions, or even trillions of individual stars.[1] They are found in a great variety of shapes, sizes, and morphologies. While many are spirals, others are round, elliptical, or irregular in shape. Galaxies can be broadly subdivided into two categories: those still forming new stars, and those which have exhausted their raw material for star formation.

Galaxies in which star formation stopped a long time ago contain only older stars. They lack the luminous blue stars, ionized nebulae, and dusty molecular clouds associated with stellar birth. They have redder colors than star-forming systems and relatively simple morphologies. Most non-star

Figure 15.1. The edge-on galaxy NGC4565 showing its prominent dust lanes and gentle warp. (J.-C. Cuillandre, CFHT).

forming galaxies look smooth and have round or elliptical shapes. Hence, members of this class are referred to as *elliptical* galaxies.

Elliptical galaxies come in a vast array of sizes. Some dwarf ellipticals consist of highly dispersed, ancient stars comparable in number to a large globular cluster. In these systems, star formation stopped soon after the very first generation of stars was born. These systems have such weak gravitational fields that the stellar winds, supernovae, and superbubbles produced by the first OB associations ejected their interstellar gas and dust. Thus they were stripped of the raw material required to sustain ongoing star formation. Though faint and therefore hard to find, dwarf ellipticals may be the most common type of galaxy in the Universe.

At the other end of the mass scale are the giant *central dominant* elliptical galaxies. Found at the centers of rich clusters of galaxies, these monster systems are the remnants of dozens or even hundreds of galaxies which merged into one. They owe their existence to *galactic cannibalism*. Like other elliptical galaxies, their light is dominated by long-lived low-mass stars. Though some of these giant ellipticals have over a hundred times the mass of the Milky Way, more than enough to retain an interstellar medium, their gas and dust

Figure 15.2. The 2 micron all-sky survey (2MASS) view of our Milky Way as seen from our location inside our Galaxy. (NASA/IPAC/Caltech).

were stripped during violent interactions with other galaxies. Therefore, they too lost the raw materials needed to sustain ongoing star formation.

The galaxies in which star formation continues to this day are lit by the blazing light of massive stars, nebulae, and OB associations. Their dusty molecular clouds can often be seen in silhouette against the backdrop of their own stars (Figure 15.1). In galaxies with little or no rotation, clouds and star-forming regions are distributed in a chaotic jumble. Such systems are known as *irregular* galaxies. But, rotation can be a great organizer. In galaxies with spin, the stars and clouds are confined to a thin rotating disk, their star-forming regions drawn into graceful arcs and spirals that wrap around their centers of rotation.

Star formation in spiral galaxies

Observations indicate that our Milky Way is a spiral galaxy. From our vantage point within the Galaxy's disk we cannot directly see the spiral arms (Figure 15.2). But the distribution of stars and gas in the sky makes it clear that the Milky Way is a highly flattened system. At visual wavelengths, interstellar dust blocks our view in many directions. For example, the "Great Rift" clouds that arch across the summer Milky Way from Sagittarius to Cygnus lie within a thousand light years of our Sun, and they block our view of the more distant regions of our Milky Way (Figure 1.1). However, along some other lines of sight, we can see much further. Careful measurements of the distances to star forming regions and molecular clouds indicate that most are confined to a pattern of spiral arms. The Milky Way's spiral arms are most obvious towards the outer parts of the Galaxy where the so-called Perseus arm can be traced for tens of thousands of light years.

Radio astronomy has revolutionized our understanding of the shape and structure of the Milky Way. Unimpeded by interstellar dust, radio waves can be used to trace the distribution and radial velocity of atomic hydrogen and a variety of molecules such as carbon monoxide. By modeling the rotation of the Milky Way, we can reconstruct an approximation of the appearance of our Galaxy to a hypothetical outside observer.[2]

Studies at optical, infrared, and radio wavelengths have shown that our Milky Way system is about 100 000 light years in diameter. Its flattened disk rotates with a speed of about 220 km/s about the center, located 30 000 light years from us towards the constellation of Sagittarius. Unlike the orbital speeds of the planets in our Solar system, which decrease with increasing distance from the Sun, the orbital speed of stars and gas remains nearly constant from the inner few hundred light years to the far outer edges of the Galaxy. However, astronomers do not find enough stars and gas in the Milky Way to produce the large orbital speeds measured in the outer Galaxy. This feature of the Galaxy provides one indirect piece of evidence for the existence of an additional mysterious substance called *dark matter*.[3]

The innermost ten thousand light year diameter region that surrounds the Galactic center (see next section) contains an elongated swarm of ancient stars that resembles a small elliptical galaxy. This is the so-called Galactic bulge. Although the inner few thousand light year region contains the greatest concentration of molecular clouds, star-forming regions, and young stars anywhere in the Galaxy, the bulge itself is relatively devoid of gas, and consists of mostly old, low-mass stars.

The bulge is surrounded by the Galactic disk and halo. Most of the Galaxy's interstellar medium, and the majority of its young stars, are concentrated in the flattened disk. A great ring of atomic hydrogen and molecular clouds occupies the region between about 10 000 and 25 000 light years from the Galactic center. This *Molecular Ring* contains several billion solar masses of gas. In contrast, the Galactic center contains about a hundred million solar masses of interstellar medium. Our Sun is located just outside the Molecular Ring. Beyond the Sun's orbit, the surface density of both gas and stars declines.

In the outer Galaxy, beyond the Sun's orbit about the Galactic center, long spiral arms are apparent. The space between these arms is devoid of molecular clouds and star formation. However, the spiral pattern is less obvious near and inside the Sun's orbit. Nevertheless, astronomers think that they can identify three or four spiral arms between us and the center of our Milky Way. But, unlike in the outer galaxy, the molecular clouds interior to the Sun's orbit populate both the suspected spiral arms and the inter-arm regions.

The swirling spiral patterns in spiral galaxies are made visible by the luminous young stars in OB associations, their associated ionized nebulae, and the dusty molecular clouds from which they form. Thus, it is star formation which renders spiral arms so visible (Figure 15.3). Why is star formation concentrated into these graceful patterns? Three different theories have been proposed and each may apply in different circumstances.

First, spectacular spiral patterns can be created by stirring from the inside by a stellar "bar," or from the outside by the passage of another galaxy. Like a spoon which produces swirls of cream in a cup of coffee, spiral patterns can be created by a rotating oblong distribution of old stars in a galaxy's center. The spiral patterns in the sub-class of galaxies known as *barred spirals* probably owe their existence to this process. Alternatively, the tides created by the close passage of another galaxy can cause stars and gas to temporarily crowd together. As these concentrations are sheared by differential rotation, they tend to be drawn into spiral arms. The resulting spiral patterns frequently wrap completely around the galaxy in a grand design.

Second, the gravity of a local concentration of matter can, under some conditions, produce a self-sustaining spiral pattern in a rotating disk. The gravitational force of such a concentration will cause all matter, both stars and gas, to migrate towards it. Thus, the mass concentration forms a wave. As in ocean waves or in a river flowing over a dip in the riverbed, stars and gas flow into the region of enhanced density, contribute to the density and gravitational field of the wave, and then leave. Such self-propagating waves can wrap much of the way around a galaxy. Matter which has come in previously, lingered, and then left, creates the gravity well which traps matter coming in later. Such self-sustaining waves are known as "spiral density waves." By concentrating interstellar gas, they locally enhance the probability that giant molecular clouds form. And where such clouds form, stars are inevitably born.

Third, self-propagating star formation discussed in previous chapters, even without the amplification provided by gravity, can naturally produce spiral patterns. Galactic disks do *not* rotate like rigid bodies such as spinning plates. Objects located at different distances from the system's center slip past each other as they orbit the galaxy. In this regard, stars and clouds in a galactic disk behave somewhat like the planets in our Solar System. Those located closer to the center make the journey about the center in a shorter time while those farther out take longer. Such differential rotation has important consequences for superbubbles and their supershells: they become elongated by the shear. Imagine looking down from above the disk of a galaxy onto an expanding supershell. Initially, the rim of the shell may look approximately round. But as it ages and decelerates, galactic shear will advance the parts closer to the galactic center, and retard the parts located further out. Looking down on a galactic disk, growing supershells become oblong ellipses. New clouds forming from them are drawn into long filaments and arcs that line up along the direction of galactic spin. Though the resulting patterns tend not to make long arms that wrap completely around a galaxy, they nevertheless produce a clear spiral pattern.

Galaxies with major spiral arms are known as *grand design* spirals. They frequently have companion galaxies which may have triggered the formation of spiral arms from the outside. Other spectacular spirals have oblong bulges that have apparently triggered spiral arm formation from the inside. If they are sufficiently massive, large-scale spiral arms can be self-sustaining. They may continue to propagate around the galaxy in the form of spiral density waves. Finally, self-propagating star formation and superbubble evolution in a shearing galaxy will tend to produce many short spiral arm segments: a configuration called a *flocculent spiral* (Figure 15.4). All three mechanisms can operate to some extent in a galaxy. For example, spiral density waves may enhance superbubble-induced cloud formation within spiral arms, and attenuate this process in the inter-arm regions (Figure 15.5).

Which of these three models is most likely to explain the recent history of star formation near the Sun? Although our Milky Way lacks a companion galaxy of sufficient mass to excite a *grand design* spiral pattern, it does possess a small bar, a cigar-shaped distribution of stars inside the inner boundary of the Molecular Ring. This bar may excite a spiral "density wave" which is most pronounced close to the Galactic center and in the outer reaches of our Galaxy. Indeed, the inner edge of the Molecular Ring contains a feature called the "3 kiloparsec arm" which can be traced nearly half way around the inner Galaxy. Along the line of sight towards the Galactic center, the 3 kiloparsec arm is approaching us with a speed of about 60 km/s. It appears

Figure 15.4. The nearby small galaxy M33 is seen almost face-on and exhibits prominent spiral arms. (T.A. Rector, NOAO/AURA/NSF).

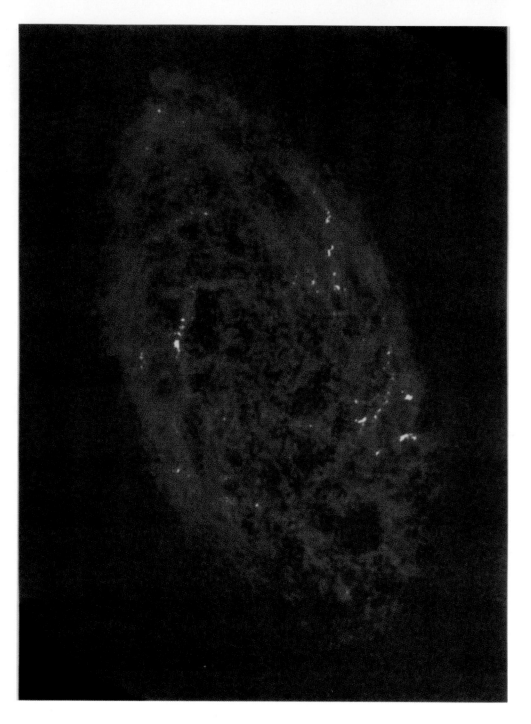

that the central bar is forcing stars and gas to "slosh" back and forth as it spins about the center. The stellar bar may not only excite spiral waves in the Molecular Ring, but it may also be responsible for producing a giant doughnut-shaped gap in the interstellar medium inside the Ring.

As discussed in Chapter 14, there is evidence that the current associations of young stars near us and their parent clouds formed from the gravity-induced fragmentation of an ancient supershell powered by a long-dead OB association. Perhaps this fossil association of stars was formed from an ancient cloud that itself was created by the last passage of a major spiral arm in our Milky Way, about 60 to 100 million years ago. Ever since, the propagating star formation discussed in the last chapter has been operating. However, there is some evidence that this process is declining. The current OB associations forming near the Sun appear to be smaller and less populated with young stars than the associations found within the major spiral arms. Perhaps, as the local interstellar medium and the Solar system move into the inter-arm region, the efficiency with which new molecular clouds form declines. Thus, the great spiral patterns in galaxies may modulate the star and cloud formation cycle associated with superbubbles and supershells.

The center of our Milky Way

The Galactic center region is hidden by massive molecular clouds, making it invisible at optical wavelengths. What we know about the Galactic center is due to infrared, radio, and X-ray observations (Figure 15.6).

Between about 1500 and 10 000 light years out from the Galactic center, there is a doughnut shaped gap in the distribution of interstellar gas. Several mechanisms may have contributed to the formation of this gap. Gas shed by planetary nebulae formed during the death of ancient low-mass stars in the bulge will interact with the orbiting gas in the disk. But the bulge stars tend on average to orbit the Galactic center on highly eccentric orbits while gas in the disk is moving on nearly circular orbits. Therefore, interaction with gas shed by bulge stars can drain angular momentum from gas originally in this region, forcing it to spiral into the Galactic center. Additionally, as discussed above, the elongated shape of the bulge (the bar) can force radial motions in the gas. Collisions with other clouds can also remove orbital energy. Observations show that the central 1500 light years of the Milky Way inside the doughnut-shaped gap in the interstellar medium, contains the greatest concentration of molecular clouds anywhere in the Galaxy. Perhaps the gas that was located between 1500 and 10 000 light years from the center ended up here. Some of these Galactic center clouds support the most energetic bursts of star formation activity in the Galaxy.

The star forming complexes known as Sagittarius B1 and B2 would each dwarf Orion's high-mass star forming regions. Their parent molecular clouds contain more than a million times the mass of the Sun and the luminosities of these Galactic center star formation regions exceed that of Orion by a hundredfold. Each of these regions is in the process of producing dense clusters containing hundreds of high-mass stars. In contrast, the Orion Nebula today contains only several OB stars.

The end products of similar but slightly older bursts of violent star formation activity dot the inner folds of our Galaxy. Rich clusters of high-mass stars abound in this portion of the sky. Many high-mass stars near the Galactic center are rapidly evolving to the supergiant phase of stellar evolution and are on their way to explode as supernovae. However, these clusters are hidden by a thick veil of interstellar dust that prevents even the brightest stars from being seen at visual wavelengths. We knew virtually nothing of these

Figure 15.6. A close-up of the center of our Milky Way as seen by the 2MASS survey in the near-infrared. Vast clouds of gas and dust hide the activities of the center region. (NASA/IPAC/Caltech).

amazingly rich star forming regions at the heart of our Galaxy until the advent of infrared telescopes and detectors in the 1960s and 1970s.

The Galactic center molecular clouds are quite different from those in the solar neighborhood. Galactic center clouds are more than an order of magnitude denser and more turbulent than clouds near to us. This is perhaps not surprising. Their location close to the heart of the Galaxy implies that, in order to be gravitationally bound and to survive the intense shear of differential motion about the center, they must be denser. Denser, self-gravitating clouds tend to be more turbulent. Additionally, these clouds are warmer than those near the Sun. High densities, high temperatures, and large turbulent motions imply that star formation, when it happens, will occur on a gargantuan scale in the Galactic center environment.

The innermost few light years of the Milky Way contain the most extraordinary group of massive stars in the Milky Way. Here, hundreds of thousands of old low-mass stars are packed into a single cubic light year of space. A similar volume near the Sun contains on average less than one star. In the midst of these old stars, dozens of peculiar high-mass supergiants form a dense cluster in the very core of the Galaxy. However, the formation of these objects remains a mystery. To survive the tides in the center of the Galaxy, a gravitationally bound cloud has to be a million times denser than Orion's star-forming core. One possibility is that these high-mass stars actually formed several hundred light years from the Galactic center about 10 million years ago in a dense cluster. The entire cluster may have been dragged into the center by "dynamical friction," a process we shall discuss later in this chapter. But to work, this process has to be very efficient.

Alternatively, it has been proposed that the high-mass Galactic center stars were formed by mergers of stars that collided. However, unlike in forming dense clusters where merging can be facilitated by disks or envelopes, these stars are not surrounded by dissipative media. Thus, collisions between normal stars that lead to merging are expected to be extremely rare. On the other hand, the probability of a merger can be greatly enhanced if one of the stars is a red giant or supergiant. These stars have greatly enlarged atmospheres that can help to trap and swallow either a companion star or an unrelated passing star. The ingestion of another star brings a fresh supply of hydrogen that may sustain further thermonuclear fusion reactions, thereby lengthening the star's life, in a process of *stellar rejuvenation*.

Recent observations have produced compelling evidence that the Galaxy's central cluster contains a black hole with a mass of 2.6 million Suns. A black hole is an object so dense that not even light can escape its intense gravity. But when matter falls into such an object, a portion of its rest-mass can be radiated away as pure energy. Thus, black holes can produce energy even more efficiently than the thermonuclear fires which sustain the lives of stars. Giant black holes lurking in the centers of galaxies can be detected indirectly by their effect on motions of surrounding stars and gas whose velocities increase greatly as they approach the black hole. The fast orbital

Figure 15.7. A group of interacting galaxies called "Seyfert's Sextet." Five of the galaxies are interacting and colliding. The sixth, the small spiral galaxy, is a background object. (NASA/STScI).

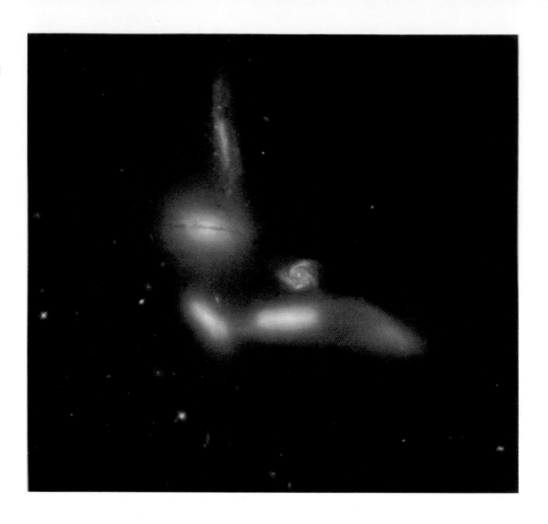

speeds of several stars in the very center of our Milky Way indicate the presence of a concentration of mass so dense that it can only be a massive black hole. Some stars in the Galactic center have been seen to move about in elliptical orbits with speeds of up to nine *thousand* kilometers per second! A faint radio star known as Sagittarius A* is thought to coincide with the black hole. But, unlike the suspected black holes in the centers of some other galaxies, the Milky Way's central black hole is currently very quiet and inactive.

We do not yet understand how black holes form in the centers of galaxies. But we do know that such black holes are common. Furthermore, black holes can be much more massive and energetic than the object residing in the center of our Galaxy. Some have masses that exceed a *billion* solar masses. And when matter falls into them, they can shine with the combined light of a thousand galaxies! Some astronomers suspect that such monster black holes are biproducts of star formation run wild.

Colliding galaxies: star formation in the extreme

Like stars in our Milky Way, galaxies also tend to cluster into groups, small and large, bound by their mutual gravity (Figure 15.7). Casual inspection of images of galaxies shows that galaxy diameters are relatively large compared to their separations: typically about 1 to 10 percent. As an example, consider the Milky Way and our nearest large galactic neighbor, the Andromeda galaxy. Each of these galaxies is about 100 000 light years in diameter. Today, they are separated by about 2 million light years. However, in rich clusters, galaxies can be packed much closer and galaxy–galaxy interactions, collisions, and mergers can be relatively common. Thus, the environment of rich galaxy clusters is hazardous to gas in galaxies. The star forming interstellar medium can be removed from a galaxy by two processes.

First, galaxy clusters are frequently filled with hot X-ray emitting plasma: a so-called *intergalactic medium*. As gas-rich galaxies move through such a cluster, their interstellar media can be violently stripped away by the ram pressure of this hot plasma.[4] In small groups, galaxies move about each other with speeds of hundreds of kilometers per second. But in rich clusters, these speeds can approach a thousand kilometers per second or more. The violent collision between the interstellar medium of a galaxy moving through the relatively tenuous intergalactic medium of a cluster can heat both to temperatures of millions of degrees. Shock-heated material is stripped away from the galaxy and eventually becomes incorporated into the hot X-ray emitting intergalactic medium.

Second, the interstellar media of merging galaxies can be consumed by high rates of star formation. The most extreme examples of star formation occur in galaxies which have just merged. Collisions between galaxies are relatively common because their typical separations are only 1 to 2 orders of magnitude larger than their sizes, especially in dense clusters and groups. The sky is littered with the debris of past galactic near-misses, head-on collisions, and galactic merger remnants.

When galaxies collide, their gas clouds smash into each other, but their stars do not. This is because, except in the very densest star clusters, the spacing of stars is so large in comparison to their sizes that they tend to just slip past each other.[5] Nevertheless, the stars feel the gravitational pull of *both* galaxies in an interacting pair. Stellar orbits become deflected by gravitational forces as the systems of stars glide past each other. Their orderly orbits about their own galactic centers become scrambled and the two star systems tend to merge into a single system (Figures 15.8 and 15.9). Stars in recently merged systems swarm around the merging nuclei from every direction, frequently in highly elongated orbits. The resulting stellar systems eventually come to resemble elliptical galaxies. Indeed, galactic merging is thought to be an important mechanism for forming giant elliptical systems, especially in the centers of rich galaxy clusters.

In a typical galactic merger, giant molecular clouds are most likely to collide with the more tenuous, low density, but pervasive interstellar medium of the other system. However, cloud–cloud collisions are also possible. Since collisions between the galactic interstellar media occur with speeds of hundreds of kilometers per second, powerful shock waves are driven into the

Figure 15.8. An interacting pair of galaxies, NGC4676, also known as "The Mice." Strong tidal forces draw out long tails of stars and gas. (NASA/STScI/HST).

clouds. When such shocks impact a molecular cloud, the medium is at first heated. But at the typical molecular cloud densities, the emission of radiation can efficiently carry away most of this energy, allowing the gas behind the shocks to be greatly compressed.

Galactic collisions produce widespread shock heating and compression and lead to the wholesale crushing of entire clouds all at once. Like a

Figure 15.9. The galaxy UGC 10214 is a spiral galaxy disturbed by a collision with a small dwarf galaxy. Blue star clusters and OB associations along the long tidal tail result from powerful bursts of star formation. (NASA/STScI/HST).

firestorm consuming a forest, powerful bursts of star formation rip through the clouds as gravitational collapse overcomes internal pressure more or less simultaneously. Since the fast shocks produced by a merger produce ultra-high pressures, such star-bursts favor the formation of massive stars and dense star clusters with very high efficiency.

The shocks produced by a galactic collision tend to synchronize star birth throughout the entire merging galaxy. The formation of ultra-dense super-star-clusters is followed by the explosions of hundreds of supernovae. When small galaxies merge, the relatively feeble gravity of the merger product may be unable to prevent the resulting superbubbles from blowing out the remaining interstellar gas into the surrounding intergalactic medium. But in large systems with stronger gravity, the interstellar medium may be retained. As the star formation cycle (the "galactic ecology" of Chapter 14) processes clouds through several generations, most of the interstellar medium is consumed and converted into stars. As discussed below, the process of "dynamical friction" tends to drag large clouds and clusters of stars into the center of the merger product. The combination of these processes can deplete the interstellar medium of a galaxy merger in less than a billion years.

Once the raw material of star formation is lost, stellar birthing stops. As the stellar populations age, massive blue stars, ionized nebulae, and luminous OB associations fade. In time, such gas-depleted spiral or irregular galaxies become indistinguishable from large elliptical systems. Thus, while spiral and irregular galaxies tend to inhabit the more isolated reaches of inter-galactic space, most large elliptical galaxies are found in the rich galaxy clusters, where collisions have been common.

Our own Milky Way may be heading towards several mergers in the far future. As mentioned earlier, the Large and Small Magellanic Clouds are spiraling into the Milky Way. Signaling their decaying orbits, a giant ring of hydrogen gas, torn from our galactic neighbors by the Milky Way's gravity, nearly encircles our Galaxy. This stream of tidal debris serves as a warning that within the next billion years, the Magellanic clouds will collide and merge with the Milky Way, perhaps triggering a starburst in our own galaxy.

A bit further lies the giant Andromeda spiral galaxy, our nearest large galactic neighbor. The Andromeda galaxy is plunging towards the Milky Way with a speed of about 200 km/s. It will take about three to four billion years to close the gap between us. A few billion years before our Sun starts to die, the Milky Way may merge with the Andromeda galaxy in the largest "train wreck" that our galactic neighborhood has ever experienced. As the remaining interstellar gas in the Milky Way and Andromeda galaxies is consumed by a major starburst, the orbits of the older stars will be scrambled, perhaps transforming the merging spiral galaxies into a giant elliptical system.

Nuclear starbursts: the power of giant black holes

Molecular clouds and large star clusters feel friction as they slowly move through the tenuous medium of interstellar gas and the background of moving stars. This is because the mass of a giant molecular cloud or star cluster draws passing stars and gas towards it. As matter flows by and is pulled in by gravity, a slight enhancement of the density of the passing matter forms in the wake region from which the cloud or cluster came. This concentration of mass exerts its own gravitational pull that in turn decelerates the moving object. This force is called *dynamical friction* since it tends to decelerate any large mass moving through a medium of stars or gas.

Within a few hundred million years, dynamical friction causes most giant molecular clouds in a galaxy merger to settle into orbits close to the nucleus of the merger remnant. Since dynamical friction also acts on large star clusters and the nuclei of the interacting galaxies, these objects too will tend to settle into the center of the merger remnant.

The accumulating gas and dust in the galactic core may produce an obscuring veil, hiding the violent shocks and interactions that occur around

Figure 15.10. The spiral galaxy M81 lies near the starburst galaxy M82 which is oriented nearly edge-on. (NOAO/NSF).

the merging nuclei. Thus, the centers of such merging galaxies frequently are invisible at visual wavelengths. As much of the interstellar medium in a merger is drawn into the core of the combined galaxy, supergiant molecular clouds form from the coalescing gas. Eventually, massive star formation ignites in the core of the merger remnant. Dwarfing the luminous star forming complexes of our Galactic center and those in the Magellanic Clouds and M33, and even the globular cluster producing bursts that may occur farther out in a merging galaxy, nuclear star forming regions can produce stars dozens of times faster than our entire Milky Way.

The strong shear in the rotating nuclear region of a galaxy tends to disrupt even the most tightly bound clusters in tens to hundreds of millions of years. Thus, while starbursts in the outer reaches of a galaxy may leave behind bound star systems, those near the nucleus do not. The dispersal of nuclear star clusters contributes to the building of galactic bulges.

The trio of bright galaxies in the Big Dipper known as M81, M82, and NGC 3077 show many signs of a close interaction (Figure 15.10). A nuclear starburst occurred within the last 100 million years in the M82 galaxy which

Figure 15.11. The M81/M82 system seen in Figure 15.10 has suffered a recent three-way encounter with another galaxy, NGC 3077. The interaction has stripped atomic hydrogen from the larger galaxy, M81, and dumped it onto M82, triggering a nuclear starburst. This radio map shows 21 cm emission from neutral hydrogen clouds. (NRAO/NSF).

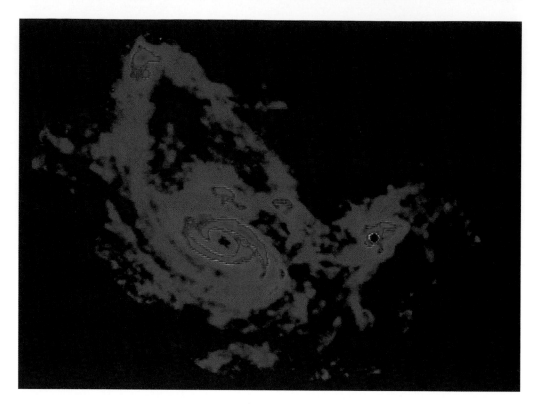

is located about 10 million light years from us. The three-way near-collision between the galaxies M81, M82, and NGC 3077 resulted in the transfer of billions of solar masses of interstellar gas onto M82 (Figure 15.11). This system has so far avoided merging, but just the long-range gravitational interaction from a close interaction had a profound effect on this group of galaxies. Tidal forces have excited a grand-design spiral pattern in M81 and generated a tidal bridge of intergalactic atomic hydrogen extending from the outskirts of M81 to M82 and enveloping NGC3077. The interaction may even have ejected several star-forming complexes from the outskirts of M81. These regions appear to be evolving into tiny dwarf galaxies whose birth is a direct consequence of the interaction. But the most spectacular consequence of the interaction is M82's nuclear starburst.

Over the last 100 million years, infalling gas has formed a ring of molecular superclouds around the nucleus of M82. Dozens of luminous giant star clusters formed within the inner 1000 light year region of the galaxy. Unfortunately, M82 is edge-on as seen from our vantage point and dust obscures the central region of this galaxy. Thus, we can only study this starburst at infrared, radio, and hard X-ray wavelengths that penetrate the veil of interstellar dust. These observations show that supernovae are exploding here at a rate 10 times higher than in the Milky Way. Thousands of recent supernova remnants have merged into a huge superbubble that is pumping shock-heated stellar ejecta and swept-up interstellar gas far above and below the plane of M82. This *nuclear superwind* has been traced out to 30 000 light years from the midplane of M82.

Over the course of hundreds of millions of years, violent nuclear star formation will consume most of the available interstellar gas. The combined effects of thousands of massive stars, their radiation, winds, and supernova explosions will expel the rest of the gas from the galaxy in a powerful galactic superwind blowing from the starburst region.

Nature is again seen to repeat herself. Galactic disks and nuclear super-winds exhibit a disk/bipolar outflow morphology similar to those seen around young stars. But instead of light-year scale flows, nuclear starbursts drive outflows that extend into inter-galactic space for tens of thousands of light years. They are one of the mechanisms for distributing the heavy elements forged in the cores of massive stars far and wide in the Universe.

Nuclear starbursts can produce hundreds of millions of young stars over the course of a few hundred million years. When combined with the older stars released by the dissolution of star clusters dragged in by dynamical friction, the nuclear regions of galaxies can attain densities of hundreds of thousands of stars within *each* cubic light year. In the company of long-lived low-mass stars, thousands of high-mass stars form and live their short but brilliant lives. Their supernova explosions leave behind ultra-dense neutron stars and stellar-mass black holes. Thus, nuclear star clusters inherit a range of objects from red dwarf stars to black holes with masses of tens of solar masses.

As a nuclear star cluster and its content of stars, black holes, and ultra-dense super-clouds interact, dynamical friction drives the system to ever increasing density. Aided by the swollen envelopes of supergiant stars, the dissipative circumstellar disks and envelopes, and ultra-dense clouds, young stars, old stars, white dwarfs, neutron stars, and even black holes become so densely packed that merging on a colossal scale becomes inevitable. The end result may be a supermassive black hole. There may be an intimate connection between the most virulent type of star formation exemplified by a nuclear starburst and the formation of giant black holes in galactic nuclei. An excellent example of a supermassive black hole is found in the galaxy M106.[6]

Matter falling into such monster black holes may power the most luminous sources of energy in the Universe, the *quasars*. Unlike stars, which are sustained by the thermonuclear burning of hydrogen into helium, quasars obtain their energy by consuming the mass that falls into them. The conversion of mass into pure energy by a black hole can be far more efficient than the conversion of one element into another. When matter falls into a monster black hole in the heart of a galaxy, its light can outshine a thousand galaxies. Due to their great brilliance, quasars are the most distant objects visible in our Universe (except for the cosmic microwave background). Some are so distant that their light was emitted when the Universe was only about 10 percent of its current age. Their properties give us a glimpse of the conditions that existed when the Universe was young.

Figure 15.12. A major jet is emerging from the center of the giant elliptical galaxy M87, probably because of material accreted by a gigantic black hole at the center of M87. (NASA/STScI/HST).

Extragalactic jets: analogs of protostellar jets?

We do not yet understand all possible ways quasars can form. Or for that matter, the remarkable diversity of processes that occur within the centers of galaxies. But a universal pattern is emerging. Wherever rotating matter accretes onto a central object, an accretion disk is formed. Spinning disks and their central objects power bipolar jets and outflows with speeds comparable to the gravitational escape speed from the point of origin of the ejecta. When the central object in the accretion disk is a protostar, the jets have speeds of hundreds of kilometers per second. When the central object is a forming white dwarf in a planetary nebula, the jets have speeds of hundreds to thousands of kilometers per second. When the central object is a neutron star, the outflows move at a fair fraction of the speed of light. When the central object is a black hole, the jets move very close to the speed of light. As the outflows interact with their surroundings, they decelerate and sweep up dense shells. Thus, we have bipolar molecular outflows, bipolar winds in planetary nebulae, bipolar nuclear superwinds, and relativistic jets from black holes in galactic nuclei. In all types of outflow, the launch is the result of a combination of thermal motion, radiation pressure, and magnetic fields.

In some ways, quasars, radio galaxies, the "Seyfert" galaxies, and other forms of active galactic nuclei[7] look remarkably similar to the young stars, their accretion disks, and jets. Though there are large differences in scale, speed, and energy, there are many similarities in other properties. The jets powered by galaxies often show multiple bow shocks and knots (Figure 15.12). Though we observe them with very different methods, their structures are superficially similar to jets produced by young stars. The powerful jets produced by active galactic nuclei show the same sorts of bends and wiggles as are seen in the flows from young stars. The sources of both types of outflow are surrounded by disks. In both classes of objects the processes visible to an observer depend both on the properties of the disk, and on its orientation with respect to our vantage point. For example, a disk seen edge-on obscures the central source in both young stars and active galactic nuclei. On the other hand, when the disk is face-on, the central object is often so bright that it obliterates the light produced by its outflow. Both types of objects ultimately derive their energy from gravity. Young stars and active galactic nuclei produce their power by tapping into the energy released by matter falling onto a central object.

Galaxy-scale disks and jets are vast, some reaching dimensions of *millions* of light years. It is not possible to observe their evolution on a human time-scale; yet, in accreting young stars changes can be seen in hours, days, or years. Thus, young stars provide an easily observed environment in which we may gain insights which can be applied to the much larger scale systems inhabiting the centers of galaxies.

16 The first stars and galaxies

Peering deeper and deeper into space, we see ever further back in time. The most distant galaxies and quasars produced their light well before our Sun and its solar system formed. This light was emitted when the cosmos was less than a few billion years old and has been traveling towards us for nearly 90 percent of the age of the Universe.

The cosmic microwave background

In 1965, radio astronomers detected an even more ancient light which probes cosmic history: the cosmic microwave background (Chapter 1). This radiation has three remarkable features. First, it comes to us from every direction. Second, it is smooth to about one part per hundred thousand. Third, the microwave background is perfectly described by what physicists call a "black-body spectrum"[1] with a temperature of about 2.7 degrees above absolute zero.

The cosmic microwave background was emitted by plasma with a temperature of about 3000K that permeated the cosmos when it was less than 300 000 years old. The radiation was emitted with a black-body spectrum which peaked at a wavelength of about a micron. However, the expansion of the Universe stretched (redshifted) the wavelengths a thousand-fold. Today, the cosmic background is visible at wavelengths near a millimeter.

Like the surface of the Sun, the plasma emerging from the Big Bang was hot and opaque. The background radiation forms a glowing barrier that prevents our observations from probing more distant space or ancient time.[2] Thus, the cosmic microwave background is a backdrop lying behind everything else in the visible Universe. Its properties provide a snapshot of the state of matter well before the existence of planets, stars, and galaxies.

As the Universe expanded and cooled, electrons began to recombine with nuclei to form atomic hydrogen and helium. Because such a gas is transparent, recombination effectively released the radiation from the previously opaque plasma. As a result, the black-body radiation started its 13.7 billion year journey towards us.

The absence of large brightness variations in the cosmic microwave background indicates that the density of matter at the time of recombination and decoupling was nearly uniform. Apparently, the pressure exerted by the radiation[3] trapped in the earlier plasma epoch prevented gravity from producing significant condensations.

The first clouds

How did the first stars and galaxies originate from the homogeneous and smooth distribution of matter that emerged from the Big Bang? When inspected in great detail, the cosmic microwave background does reveal subtle blemishes. The primitive hydrogen/helium plasma was ever so slightly denser in some places and more tenuous in others. These density fluctuations formed around concentrations of dark matter.

Without dark matter, galaxies and stars would have been hard to form from the amount of ordinary matter present in the Universe. The formation of self-gravitating structures from the extremely smooth gas distribution prior to recombination requires the additional gravitational attraction of dark matter. Computer models show that dark matter started to form self-gravitating clumps with masses similar to dwarf galaxies well before recombination. So long as the cosmic plasma remained ionized, radiation pressure kept ordinary matter from falling into these pits. However, once the plasma recombined and the gas lost its support, it fell into the concentrations of dark matter (Figure 16.1). Within a few hundred million years, dense clouds of gas accumulated and collapsed to form the first stars.

To collapse under the influence of its own gravity, the escape speed from the surface a cloud must exceed the motions of its particles. At the 10K temperatures and average densities of modern day molecular clouds (Chapter 3), regions containing a little less than one Solar mass have enough mass to overcome the internal motions. In contrast, the primordial clouds of nearly pure hydrogen and helium had temperatures of thousands of degrees. The smallest clouds that could collapse required nearly a million Solar masses (about the mass of globular clusters and modern giant molecular clouds).

The continued collapse of a cloud requires that energy liberated by gravity be radiated away. Collapse increases the energy of atomic and molecular collisions. In modern clouds, molecules such as carbon monoxide radiate this energy as millimeter and sub-millimeter wavelength spectral lines. They act as coolants for the collapsing cloud. Additionally, collisions with dust grains result in production of far-infrared continuum radiation. None of these processes operated in primordial clouds: there were no molecules and no dust grains to emit far-infrared and sub-millimeter radiation. Thus, the first generation of clouds after the era of recombination must have remained very hot and completely transparent.

In present-day clouds, molecular hydrogen is created efficiently on grain surfaces. But in the absence of dust, molecular hydrogen can only form by means of inefficient reactions among atoms. Models of primordial clouds

Figure 16.1. A numerical simulation showing the filamentary distribution of dark matter. Galaxies will eventually form within the complex network of dark matter structures. The region shown is a billion light years across. (Ben Moore).

show that their H_2 abundance was well below 1 percent. Thus, primordial clouds remained predominantly atomic with only a trace of molecular hydrogen.

Atomic collisions at temperatures below 10 000 K cannot excite even the lowest energy levels in atomic hydrogen or helium. Thus, molecular hydrogen was the only available cloud coolant. Molecular hydrogen has a variety of near- and mid-infrared wavelength transitions, which in modern, much colder, star-forming clouds are excited only in outflows from young stars (Chapter 6). These same spectral lines radiated the energy liberated by gravitational contraction of the hot primordial clouds.

Following recombination, there were no sources of light in the early Universe. Thus, this era is referred to as the cosmic "Dark Age." As the first star-forming clouds evolved, their feeble molecular hydrogen radiation provided the first hints of the starlight about to illuminate the cosmos.

The first stars

What kind of stars did these translucent, warm, and atomic clouds produce? By extrapolating our understanding of present-day star formation, astronomers think that all the first stars were very massive. Computer models of "metal-free" clouds indicate that fragmentation is difficult, so a cloud would not divide into many small cores. As the very first star ignited in the densest part of a giant cloud it ionized its surroundings and inhibited the birth of other stars nearby, thus keeping the surrounding reservoir of gas for itself. Therefore, these first stars may have accumulated tens to hundreds of times the mass of the Sun.

Primordial clouds lacked magnetic fields. Thus, any primordial disks would be transparent, hot, and non-magnetized; angular momentum could not be removed by the magnetic processes thought to operate in modern disks. Therefore, the first stars may have been fast rotators. In the absence of dust, disks could not form planetesimals or planets: the raw material for planetary system formation was absent.

Modern-day protostars produce their first radiation at sub-millimeter wavelengths. Their spectra gradually shift from the sub-millimeter to the visual portions of the spectrum over the course of hundreds of thousands of years. But not so for the very first stars. Composed of pure hydrogen and helium, they started their lives by emitting visible and near-infrared radiation right from the beginning.

Primordial clouds were heated by their collapse as the burden of infalling gas raised internal pressures and temperatures. As molecular hydrogen dissociated, the total absence of coolants allowed temperatures to climb until hydrogen atoms were re-ionized by collisions. Free electrons once again generated light. When the first stars emerged about 400 million years after the Big Bang, the cosmic background radiation bathed the Universe with mid-infrared light. The feeble glow of molecular hydrogen produced by collapsing clouds was soon followed by a growing red and near-infrared glow from the plasma of massive protostars.

These spheres of ionized hydrogen/helium plasma were opaque to their own radiation, with surfaces delineated by the transition from atoms to ions. Thus, free electrons in the plasma could radiate to space. As these first protostars grew, so did the area of their radiating surfaces and they became brighter. Though briefly resembling red giants when born, they rapidly evolved into luminous, blue, and very hot stars.

As a protostar shines, its core shrinks to ever greater density, pressure, and temperature, until eventually thermonuclear fusion of hydrogen ignites and replaces gravitational contraction as the primary source of stellar energy.

Modern-day massive stars burn hydrogen to helium by means of the so-called carbon-nitrogen-oxygen cycle. But in the early Universe there was no carbon, so another process based only on protons would operate. However, this path to helium requires much greater central pressure and a higher temperature to sustain the prodigious luminosity of massive stars. As a result of this, massive first-generation stars produced far more ionizing ultraviolet light at their surfaces than modern-day stars of comparable mass. Even stars with masses of only a few Suns would have been considered O or B stars. Because of this copious production of ultraviolet radiation, the translucent surroundings of primordial stars were re-ionized. By the time the most distant galaxies we see today were formed, most of the gas in the Universe was reheated and ionized a second time (the first time being when it emerged from the hot Big Bang). The tenuous and transparent medium between the galaxies remains ionized to this day.

Because the first generation of stars were massive, they were short lived. They synthesized the first heavy elements in their cores and then evolved toward an explosive death. Since these stars contained no metals (elements heavier than helium), stellar winds were either very weak or absent.[4] Thus, when they exploded, their masses were greater than those of modern massive stars.

The first clouds were super-massive and could not fragment as easily as today's clouds. Thus, rather than forming in very dense clusters, the first stars were likely born in looser clusters and associations. Their collective hard-UV radiation fields ionized giant cavities in the infalling envelopes. Hundreds of supernovae marked the deaths of these stars.

The products of thermonuclear reactions were ejected and mixed with surrounding gas. Element-enriched superbubbles expanded and collided with infalling hydrogen in its rush towards the local gravity well. While in some places the bubble walls burst into more tenuous intergalactic space, at other locations shocks compressed the infalling gas and induced second generation star formation. These clouds became somewhat enriched with the heavy elements produced by the first stars, leading to more efficient cooling, fragmentation, and lower-mass cloud cores which produced lower-mass stars. As heavy-element abundances slowly increased, the typical masses of forming stars gradually decreased.

Initially, the young Universe was made entirely of interstellar and intergalactic gas, the raw material of star birth. There was more hydrogen and helium available to fuel star formation than at any time since. The first generation of star formation must have ignited nearly simultaneously in many locations. These regions may have resembled gas-rich dwarf irregular galaxies we see in the sky today. But there was one important difference: these early dwarf galaxies were dominated by warm gas which could only produce massive stars.

Since we have not yet detected the light produced by the first stars, we do not fully understand the history of star formation in the Universe. Observations of the cosmic microwave background taken together with studies of the evolution of galaxies indicate that the very first stars formed no more than 400 million years after the Big Bang. However, in some under-dense portions of the cosmos, star formation may not have ignited till much later, perhaps not before several billion additional years had elapsed. Astronomers think that most primordial ("metal"-free) star formation occurred within the first 10–20 percent of the current age of the Universe. Studies of distant galaxies and the fossil record of old stars within our own Milky Way show that the average cosmic rate of star formation peaked about 3–8 billion years after the Big Bang, well before the formation of our Solar system. By the time the Sun formed 4.5 billion years ago, most of our Galaxy's interstellar medium had been converted into stars. Ever since, the average rate of star formation in the Universe has been declining as galaxies converted more and more of their remaining interstellar matter into low-mass long-lived stars.

The formation and evolution of galaxies

Clusters of first generation massive (or supermassive) stars exploded and polluted their growing superbubbles with freshly made heavier elements. Elements such as carbon, silicon, and iron started to form dust grains and simple molecules. Thus, the primordial high-mass stars created the conditions in which dusty and opaque molecular clouds resembling those of today could take shape. As the first small galaxies assembled, their molecular clouds would soon support star formation much as we know it today.

As we look deeper into space, and further back in time, we see galaxies and stars receding from us at faster and faster speeds. The first stars were thought to form so long ago that the expansion of the Universe shifts their light from the UV and visual spectral region into the infrared. We have to search the farthest reaches of space at near-infrared wavelengths to glimpse the first starlight. NASA's 6 meter *James Webb Space Telescope* is designed to search for this first light.

As freshly synthesized elements condensed into dust grains in the ejecta of dying massive stars, the primordial gas became dusty and started to re-process light into longer-wavelength radiation in the far-infrared and millimeter portions of the spectrum. To detect the radiation re-emitted by the first dust clouds, we have to search the radio part of the spectrum. Astronomers are upgrading the VLA and contemplating giant space-based sub-millimeter telescopes to search for this re-processed radiation.

During the first few billion years of cosmic evolution, the density of matter was much higher and the spacing between forming galaxies was much smaller. Interactions between neighboring regions were common. The first galaxies suffered mergers that lead to the buildup of ever larger systems of dark matter, stars, and gas.

These first galaxies are thought to have been similar to today's dwarf irregular systems which have relatively small mass compared to galaxies such as the Milky Way. In dwarf systems, stars and gas clouds are bound only weakly by gravity. When clusters of massive stars form and die, the resulting supernova explosions and superbubbles can flush most of the remaining interstellar gas completely out of the system, bringing star formation to a halt. Such systems were left with relatively few stars formed only during the first waves of star formation.

Galaxies with sufficient mass to retain their interstellar matter following the expansion of the first superbubbles produced wave after wave of star formation. As clouds were contaminated with heavy elements, molecules and dust formed, and the gas could cool more efficiently. Collapsing clouds would fragment into smaller, less massive clumps, which would form lower mass stars. These stars still had very low abundances of elements heavier than helium. Astronomers have found several low-mass stars in the halo of the Milky Way that contain less than one ten-thousandth of the metal content of the Sun. These metal-poor halo stars may be the most ancient remnants from the era of galaxy formation.

As more and more hydrogen accumulated in the dark matter halos and became more metal-enriched, star formation evolved towards its modern state. Populations of long-lived low-mass stars developed and grew in the centers of small proto-galaxies. These systems interacted violently with their siblings in a chain of galaxy collisions that combined their young stellar populations. Gas compressed by such collisions promoted star-bursts in their interstellar media.

Like the growth of planets from proto-planets, dwarf galaxies merged into larger systems. These early mergers probably built some of the spheroidal bulges that would become the backbones of today's spiral galaxies (see Chapter 15). Eventually, hundreds of dwarf galaxies would merge into large modern systems. As most of the primordial metal-free gas rained onto the merging galaxies, high angular momentum gas formed spinning disks and eventually evolved into the spiral systems such as the Milky Way and Andromeda galaxies.

Deep images show that, at high redshifts, the distant Universe contained mostly dwarf irregular galaxies with abundant star formation, some disk galaxies similar to the Milky Way, and relatively few elliptical systems. Interactions and mergers were far more common in the early Universe than today and these processes played a fundamental role in the early building of galaxies and the promotion of virulent star formation.

Observations of quasars and other phenomena related to activity in galactic nuclei (see Chapter 15) demonstrate that many galaxies formed supermassive black holes at their centers within a few billion years of the Big Bang. The most distant quasars were spewing forth vast amounts of radiation and powerful jets billions of years before the formation of our Sun. Furthermore, the abundance of heavy elements became comparable to or larger than the Sun in many systems at relatively early times. Such rapid buildup of heavy elements indicates that star formation in some large galaxies must have been vigorous, consuming gas at rates orders of magnitude faster than the Milky Way consumes gas today. As we look back at the ancient quasars, their spectra are laced with the products of thermonuclear fusion reactions in massive stars.

In the most over-dense portions of the Universe, merging and galactic cannibalism consumed most gas-rich dwarf irregular and spiral galaxies, converting them into giant elliptical systems. Gravity ensured that in some locations great *clusters of galaxies* were assembled. The interstellar media of interacting and merging systems were expelled into the intergalactic void and heated to X-ray temperatures by shocks produced by galactic collisions, nuclear superwinds, and active galactic nuclei. Today, such large galaxy clusters are prolific sources of X-ray emission. They also are systems in which star formation has all but stopped. Most of their normal matter is either locked away in long-lived low-mass stars, or floating about the intergalactic cluster medium as a tenuous X-ray plasma.

Over the eons, most primordial hydrogen and helium left over from the Big Bang and the era of re-ionization fell into galaxies, forming the raw material for star formation. In elliptical galaxies, star formation came to an end as interstellar gas was consumed by star formation or expelled. In other galaxies, huge starbursts produced active galactic nuclei and in this way exhausted their supply of gas. And in still others, such as the Milky Way, star formation continues to this day, because they formed far from rich galaxy clusters and therefore retained enough interstellar gas to produce new stars.

In all of these diverse environments, the star formation cycle that we are familiar with in the Solar neighborhood started once the metallicity became high enough: from molecular clouds, to stellar birth, stellar death and explosion, the expansion of superbubbles, the sweeping up of giant shells, and the condensation into new clouds. With each cycle, a portion of the remaining interstellar matter was removed from space and incorporated into low-mass long-lived stars which effectively store their contents for billions of years. The remaining interstellar gas became more and more enriched by the elements forged by the short-lived massive stars.

17 Astrobiology, origins, and SETI

Astrobiology and the origins of life

Is life on Earth unique or is life abundant in the Universe? This question has long been of great interest to scientists, philosophers, and the general public.

Star formation is a key step in the cosmic evolution of matter from the Big Bang to life as we know it. Without the thermonuclear burning of the light elements hydrogen and helium into the "metals" in the cores of massive stars, the very ingredients from which planets are made would not exist. The ingredients of life (the carbon, nitrogen, oxygen, and other elements used to build the organic molecules that are the basis of life) were forged in the centers of stars, and recycled by stellar winds and explosions into the interstellar medium from which our Sun and Solar System formed. We are literally stardust. Furthermore, long-lived low-mass stars, and the rocky planets which sometimes orbit them, provide the stable and nurturing environments in which the chemistry of life may become established and may evolve into the complex forms which can support intelligence. Star formation has played a crucial role in our emergence.

It appears that organic chemistry (the chemistry of carbon) naturally leads to great complexity. This is evident in the composition of interstellar molecular clouds from which stars and planetary systems form. Well over half of all known interstellar molecules, including most complex molecular species, are organics. Studies of meteoritic evidence demonstrate that planetesimals were replete with complex organic molecules. Thus, the primitive Earth inherited a rich chemistry. Though most of the complex molecules that went into the formation of the Earth may well have been destroyed during the late phases of bombardment, subsequent impacts by comets and other carriers of organics almost certainly delivered substantial quantities of the compounds of life to the surface of the primitive Earth.

Although this "Hadean" epoch of the Earth's early history remains poorly understood, the earliest rock records demonstrate that primitive life developed soon after the end of heavy bombardment. The Earth's fossil record shows that primitive life existed nearly 4 billion years ago, within about 500 million years after the birth of the Sun. Biologists, geologists, paleontologists, and astronomers are starting to work together in the new discipline of "astrobiology," one goal of which is to understand the origins of life here on Earth.

Since the 1970s, scientists have been searching for life on other bodies of our Solar System. Until the mid-1990s this search was entirely focused on Mars, the fourth planet from the Sun, which is more similar to Earth than any other of the Sun's planets. There is geological evidence suggesting that Mars was once endowed with a thick atmosphere and had a warmer climate. Its present-day desert surface contains channels which may have been formed billions of years ago by freely running water. Even now the poles of Mars are covered by layers of ice and frost. However, the two Viking spacecraft that landed on Mars in 1976 found no evidence of biological activity.

Large meteor impacts on Mars occasionally dislodge bits and pieces of its surface with such force that they are launched into space. A tiny fraction of such ejecta may enter the Earth's atmosphere to fall as meteorites. These incredibly rare samples of the Martian surface can be identified by the unusual mixture of argon isotopes in microscopic pockets of trapped gas. A meteorite found in 1984 in the Allen Hills region of Antarctica turned out to be from Mars. Upon microscopic examination, it was found to contain structures remarkably similar to ancient fossils produced by microbes here on Earth. The great interest generated by this finding led NASA to design a series of missions to Mars that may culminate in the return of fresh samples of the Martian surface not altered by the violence of a meteor impact, a long voyage through space, or prolonged exposure to contaminants here on Earth. With luck, such fresh Martian samples may settle the question of whether or not primitive life once existed on the Red Planet.

The underlying premise of astrobiology is that liquid water must be present for life to exist. To have liquid water, a planet must be in a stable, nearly circular orbit, at the "right" distance from its parent star. A significant portion of its surface must have a temperature above the freezing point of water. For Sun-like stars, an orbital radius not too different from that of the Earth is required. Though the terrestrial fossil record shows that life required at most 500 million years to appear, it took an additional three and a half *billion* years of evolution for large animal and plant life to transform the Earth's surface and atmosphere into today's familiar state. Life as we know it is likely confined to Earth-like rocky planets in 1 AU orbits around long-lived low-mass stars of spectral type G such as our Sun. Stars which have a factor of two more mass do not live long enough to enable biological evolution on a time-scale of 3 to 5 billion years. Stars which have a factor of two less mass, are much dimmer than the Sun. To be warm enough to enable surface water to remain liquid, a planet would have to orbit so close to such a star that its spin would be locked to the length of its year by tidal forces. This would probably cause most of its water to freeze out on the side of the planet which is permanently facing away from the star. However, this view is biased by our knowledge of the one place in the Universe where we know life exists, Earth.

Recent missions to Jupiter have alerted astrobiologists to a potentially even more exotic habitat for Solar System life than Mars: Jupiter's second moon, Europa. Though covered by a thick mantle of ice, Europa's surface lacks the craters which mar the ancient surfaces of other small Solar System bodies like the Moon. Evidently, Europa's surface is relatively young. The images returned by the Galileo spacecraft show thousands of criss-crossing cracks in Europa's surface ice. Planetary scientists reason that Europa's global glacier must be only a few to tens of kilometers thick and may float on top of an ocean of liquid water. What keeps Europa's ocean liquid? A continuous heat source is the tidal flexing produced by gravitational interactions with Jupiter and its other moons. The discovery of life-forms on Earth which survive far from sunshine on chemicals ejected by hot water vents deep down on the ocean floor has led to the suggestion that primitive life might exist in Europa's ocean, heated by the strong tidal forces of Jupiter.

Could life exist on one of the many hot Jupiters discovered around other stars? Some gas giants orbit their host stars at a distance of about an AU, well within the habitable zone. However, gas giants are not thought to possess solid surfaces beneath their thick atmospheres. Thus, with our present perspectives on the conditions required for life, gas giants are unlikely habitats. However, gas giants may be orbited by moons – and they may be more hospitable. Such satellites may have sizes similar to Earth, might support an atmosphere, and possibly oceans. These satellites may provide a possible habitat for life.

Terrestrial life is carbon based. Are there alternatives? Science fiction writers have speculated about life based on silicon, an element which, under some conditions, may also support a complex chemistry. However, current thinking suggests that silicon-based life is improbable. Observations show that carbon-based organic chemicals are far more common in interstellar molecular clouds than compounds based on any other element. Therefore, planetary systems inherit a great abundance of carbon-based organic compounds. Carbon supports a rich and highly complex chemistry at the same temperature at which water is liquid and an ideal solvent. Furthermore, the common oxides and hydrides of carbon (carbon dioxide and methane) are gases at room temperature and pressure. In contrast, the oxides of silicon are solids, and remain so at temperatures below about a thousand degrees. Carbon-based life depends on the respiration of carbon-bearing gas to obtain this basic element. In primitive planetary atmospheres, carbon is readily available in several gaseous forms. Imagine a silicon-based life-form. To obtain its raw material, it would have to breathe (or otherwise consume) a solid such as rock or sand! But, perhaps our thinking is short-sighted. Our own technology is producing complex computing machinery from silicon-based microelectronics. Might our computer technology eventually evolve into self-replicating forms of artificial life and intelligence and qualify as genuine silicon-based life?

The light of life

On planets with free oxygen and liquid water, there may be life.

Among all known satellites and planets with atmospheres, Earth alone has abundant liquid surface water and large amounts of free oxygen. Free oxygen is so highly reactive in a chemical sense that it is rapidly depleted by a variety of "oxidation reactions." Most common elements found on planetary surfaces such as silicon, aluminum, carbon, magnesium, or iron will rapidly react with oxygen to form common minerals such as quartz, aluminum oxide, or iron oxide (rust), or common substances such as carbon dioxide and water. Without an efficient process to release oxygen bound into minerals, a planet's atmosphere would rapidly lock away and hide its oxygen. The respiration of plants appears to be by far the most efficient mechanism for releasing free oxygen into an atmosphere. There are no known natural processes other than respiration which can maintain the amount of oxygen present in the Earth's atmosphere. Thus, free oxygen may be a powerful tracer of life.

The presence of active biology on a remote planet can in principle be identified from its spectrum. The presence of water would signal a potential habitat. Finding the spectral bands of molecular oxygen (O_2) or ozone (O_3) in a planet's light may be an indirect signature of biological activity. This test, if applied to all Solar System bodies, would produce a positive result only for Earth.

To apply this "oxygen test" to extrasolar planets, we must build a telescope that can isolate the light of a distant planetary system from that of its parent star. Then we will be able to apply this test for life by taking a spectrum of the planet. Fortunately, the two common molecular forms of free oxygen, molecular oxygen and ozone, have prominent spectral features in both the visual and infrared portions of the spectrum. Future large space telescopes which can detect a distant planetary system can also search for the "light of life."

Messages from the deep?

Are there *intelligent* life-forms elsewhere in the Universe? If so, is there any chance of detecting "them" or, better still, exchanging messages or even visiting them? The great distances between even the nearest stars and the finite speed of light severely limit our attempts at searches and "contact."

Nevertheless, in 1960 Frank Drake proposed to search for radio signals produced by extra-terrestrial civilizations. Ever since, a number of the world's radio telescopes have on occasion been used to search for alien signals. Presumably, these signals would look very different from the naturally occuring emission radio astronomers study. Alien signals may resemble the jumble of radio emissions produced by human utilization of radio technology, that is narrow-band radar transmissions, short duration bursts of radio energy, and the carrier waves of various television or radio broadcasts. The quest for such signals is called the *Search for Extra-Terrestrial Intelligence* or SETI.[1]

In principle, there are three types of signals to be sought: eavesdropping on interstellar communications, looking for beacons intentionally lit to draw attention to a civilization, or the unintentional "leakage" of signals which are the by-product of the general use of radio technologies. The first two possibilities imply civilizations far more advanced than ours. In the third case, however, we can search for civilizations in a comparable state of development to us.

If we use our own communications industry as an example, it is clear that eavesdropping on aliens is very hard. This industry has realized that it is very expensive to beam signals in every direction. Efficiency demands the use of highly directional signals or even "leak proof" media such as optical fibers. Furthermore, for point-to-point communications, radio engineers have realized that efficiency dictates the use of the highest possible frequency or shortest possible wavelength that propagates through the medium. It is possible that other civilizations use powerful visual-wavelength laser signals either for communications or as beacons. Thus, some SETI enthusiasts have established searches for narrow-band optical signals or pulses. Extrapolation of this principle to SETI implies that a space-faring civilization may use high energy gamma-rays to communicate with its outposts or ships. Though such conjectures try to guess what a more advanced civilization is likely to do, and hence are very risky, the implication is that radio searches are *not* likely to detect signals exchanged within or between advanced civilizations.

What about beacons intended to be detected by either others or the alien's own ships? This is unlikely. If *designed* to be detectable by our type of civilization, the beacons would have to broadcast orders of magnitude more power than the energy we radiate into space unintentionally. We "leak" some hundred million watts into space in the form of thousands of radio stations, radar signals, and other broadcasts. To emit much more would require formidable power reserves and investment of resources. Furthermore, *we* have not been motivated to establish such beacons to alert ET, if they are out there, of our presence. So, why should we expect another civilization to do so?

It therefore seems that the most promising type of signals to search for is the unintentional leakage produced by alien civilizations which have just achieved the use of radio technology, but have not yet developed leak-proof communications. Consequently, they are likely to be in a similar state of development to us. We can make a crude estimate of the most likely distance to such a civilization, and therefore the size of the radio telescope needed to detect its "leakage" signals. To make this estimate, we have to guess the lifetime of such a civilization, the number of habitable planets in our Galaxy, and the probability that, given such a planet, intelligent life actually evolves to the utilization of radio-frequency technology.

Human civilization has been using radio for only about a century. Contrast this to the nearly five *billion* year age of the Earth. A conservative assumption about the duration of such a "radio broadcasting" phase of civilization is that it lasts twice as long (perhaps about two centuries). Assuming that radio-civilizations are born at random times, the chance that we are around *at the same time* as another similar civilization on another planet is four parts per *hundred million*.

Let us *assume* that ten percent of all Sun-like stars in the Galaxy are orbited by habitable planets. There are about 10 billion Sun-like stars in the Milky Way, which according to the above assumption, implies that there may be 1 billion habitable planets. If so, everywhere we look at the night sky we are looking towards possible habitats for life (Figure 17.1). Further, assume that *all* produce a civilization at some time in their history.

Based on these assumptions, an upper bound on the number of civilizations similar to ours is given by the chance that any two are around at the same time multiplied by the number of habitable planets. By this reasoning, we find that at any one time, there may be fewer than 40 civilizations in the Galaxy which are in a similar state of evolution to us, that is broadcasting unintentional leakage radio signals.

Assume that these civilizations are spread randomly around the 100 000 light year diameter disk of our Milky Way. Therefore, the average separation between civilizations is about fifteen thousand light years! At the speed of light, a round-trip message would take thirty thousand years – three times the interval that has elapsed since the end of the last ice age! Therefore two-way communication is not likely between such civilizations.

We can estimate the size of the radio telescope needed to detect the hundred million watts that *we* unintentionally broadcast from a distance of 15 000 light years. To detect a similar signal from an alien civilization, such a telescope would have to have a collecting area of hundreds of square kilometers. The cost is likely to be similar to that of the exo-planet imager discussed at the end of Chapter 13. An even chance of success requires orders of magnitude improvements in the sizes of radio telescopes used for SETI searches.

The current SETI effort is likely to produce results only if other civilizations build powerful beacons in the hope that we, and other young civilizations, will find them. But as we have seen, improvements in telescope technology may enable the detection of Earth-like terrestrial planets within the next one or two decades. There is hope that we can, in future decades, continue to build ever larger telescopes. Perhaps in some future decade, we may be able to build an instrument large enough to pick up the faint leakage signals produced by a nascent technological civilization on a distant planet.

Finale

During the last few decades, we have learned much about the birth of stars and planets, and the conditions that led to our own origins.

Our Universe started in a hot Big Bang about 14 billion years ago. The emerging plasma of hot hydrogen and helium recombined when the

Figure 17.1. Among the multitudes of stars and planetary systems, where do we find the few that may hold intelligent life at the moment? (J.C. Cuillandre/CFHT).

Universe was 300 000 years old and is still visible to us as the cosmic microwave background. The birth of the first stars marked the end of the subsequent "Dark Age" nearly a half-billion years later. Collisions and mergers of these first galaxies, and the infall of gas left over from the Big Bang led to the emergence of the large galaxies and galaxy clusters we see in our sky today.

The primordial stars were massive. Their cores forged the chemical elements by thermonuclear fusion of hydrogen and helium. The terminal supernova explosions of many generations of such massive stars injected heavy elements into interstellar space. Enriched giant molecular cloud cores cooled and fragmented into longer-lived, lower-mass stars.

Our Solar System is about 4.5 billion years old. Recent evidence indicates that our home world formed in an OB association along with thousands of sibling stars. The environment may have stimulated the growth of solids and injected short-lived radioactive species into the forming planetary system. The sedimentation of refractory elements in spinning circumstellar disks formed rocky and icy planetesimals. Short-lived radioactivity differentiated these planetesimals and produced the chemical make-up of some primitive meteorites. Through ever more violent collisions, these asteroid-mass bodies coalesced into rocky planets and cores of the gas giants orbiting their parent stars.

The primitive Earth was pelted by the left-over debris of planetesimals, making life impossible for about half a billion years. But, as the bombardment abated 3.8 billion years ago, Earth accumulated a protective atmosphere and an ocean of water from the icy cometary material left over from its protoplanetary disk.

The "metals" from the centers of prior generations of massive stars congealed into "life" which eventually evolved into curious humans who want to understand their own origins. So the Universe becomes aware of itself. Perhaps out there among the stars, there are others who share this awareness.

Appendix   Notes to the chapters

Chapter 1

1   Spectroscopy:

A spectroscope (or spectrograph) sorts and spreads out light according to its constituent wavelengths (colors). There are two common devices which can achieve this: prisms and diffraction gratings.

A prism refracts (or bends) light by an amount that depends on the angle of incidence and the *refractive index* of the material. For most materials, the refractive index (given by the ratio of the speed of light in a vacuum divided by the speed in the medium) varies with wavelength.

Diffraction gratings consist of mirrors (reflection gratings) or transparent material (transmission gratings) with a large number of parallel grooves. Some of the light reflected (or transmitted) by a grating is bent with respect to the normal reflected (or refracted) ray by an angle which is a function of the ratio λ/D, where λ is the wavelength of light and D is the spacing of grooves.

2   Super-massive stars:

The most massive stars known have masses about 100 times that of the Sun. Some astronomers have proposed that stars might exist with considerably greater mass. Hypothetical stars with greater than about 100 times the mass of the Sun are called *super-massive stars*.

3   Ordinary matter:

Ordinary matter made of atoms appears to constitute only about 4 percent of the mass of the Universe. Observations have provided evidence that most of the Universe is composed of "dark matter" which does not absorb or emit light at any wavelengths. However, dark matter does exert and experience the gravitational force. Indeed, gravity provides compelling evidence that about 30 percent of the mass in the Universe consists of dark matter.

Evidence has recently emerged that nearly the equivalent of about 70 percent of the mass of the Universe is an even more exotic entity called "dark energy." The evidence for this takes the form of an apparent acceleration of the expansion of the Universe during the last 5 to 10 billion years. This evidence has been uncovered by the observation of Type Ia supernovae at great distances that act as "standard candles." See footnote 3 in Chapter 15 for more discussion.

But, in the developing picture of the elementary particles, the particles themselves consist of loops or "strings" trapped by so-called "topological defects" in the curvature of space-time that develops at the level of a Planck length. What are these "defects"? The handle of a coffee cup can serve as an oversimplistic example. Suppose that when you made the cup and its handle, you placed a wedding ring about the handle. After the cup has become solid, there is no way of removing the ring without either breaking the ring or the handle of the cup. Thus, certain miniature black holes can be stable. Such holes may actually represent the elementary particles of nature.

In the so-called "superstring" or "M-theories" of matter and energy that have become popular since the mid-1980s, on the scale of a Planck-length, all matter and energy consists of tiny string-like loops or membranes of space-time similar to the wedding ring trapped on the handle of the coffee cup. The vibrational modes of these tiny entities determine the physical properties of particles such as mass, charge, particle type, and the forces to which they respond or carry. Interactions between particles and particle transformations involve changes in the vibrational states of these loops and membranes. The creation or annihilation of elementary particles involves the formation of the topological defects and their undoing, like fabrication of a cup handle from a sheet of hot glass, or the melting of the cup back into a flat sheet.

An excellent popular book on this subject is Brian Greene's *The Elegant Universe* (Random House, 2000).

8   Fusion:

The critical step in the thermonuclear fusion of hydrogen to helium (the process that makes most stars shine) involves the inverse of the neutron decay process, and can only occur when the hydrogen plasma is sufficiently hot so that collisions can overcome the "energy barrier." A sufficiently energetic collision between an electron and a proton can produce a neutron and the anti-neutrino. The collision has, in effect, converted the up quark of the proton into the down quark of the neutron. If the density is high enough, the neutron will collide with a proton before the neutron decays, and become bound to it by the strong nuclear force. Since strongly bound neutrons do not decay, the resulting particle, a bound proton-neutron pair called "deuterium" (or heavy hydrogen) is stable and long-lived. Collisions between pairs of deuterium lead to the formation of ^{3}He – a helium nucleus containing two protons and one neutron. The collision of two ^{3}He nuclei leads directly to the formation of a nucleus of ^{4}He which contains two protons and two neutrons. This chain of reactions requires a temperature of about 10^7 degrees, set by the first step in the process – the formation of deuterium. As mentioned in Chapter 4, deuterium can "burn" to ^{4}He at a temperature of 10^6 K. In stars somewhat more massive than the Sun, higher core temperatures enable the conversion of H to ^{4}He by a "catalytic" process involving carbon 12 (^{12}C). In this so-called "CNO" process (named after the elements carbon, nitrogen and oxygen which participate) ^{4}He is produced by collisions with protons followed by neutron decay. $^{12}C + H \rightarrow$ $^{13}N \rightarrow {}^{13}C$; $^{13}C + H \rightarrow {}^{14}N$; $^{14}N + H \rightarrow {}^{15}O \rightarrow {}^{15}N$; $^{15}N + H \rightarrow {}^{12}C \rightarrow {}^{4}He$. See also note 1 to Chapter 14.

9 Temperature scales:

Physicists measure temperature on the Celsius (or Centigrade – abbreviated as C) scale in which water freezes at 0° C and boils at 100° C. Absolute zero, the temperature at which the random motions of atoms stops, corresponds to −273° C. The Kelvin (abbreviated as K) temperature scale is defined by making −273° C equal to 0 K. On the Kelvin scale, water freezes at +273 K and boils at +373 K.

10 Cosmic expansion:

The expansion of the Universe occurs in three-dimensional space, whereas a balloon has a two-dimensional surface. Thus, this analogy only works if the balloon has a three-dimensional surface, which is the same as a set of nested concentric balloons. Imagine a set of nested balloons, each of which is inflating so that the ratios of the radii of the balloons remain constant. Galaxies "painted" on the surfaces of these nested balloons behave like the galaxies in the Universe if the rate of radius increase of each balloon is constant in time.

11 Ionization:

When the number of electrons does not equal the number of protons for a given element, it is called an *ion*.

The removal of electrons from an electrically neutral atom leave behind a positively charged ion. Radiation and energetic collisions are both capable of stripping electrons from atoms and ions. The addition of electrons to an electrically neutral atom produces a negatively charged ion. In most astronomical settings, atoms are either neutral or have lost one or more electrons. Positively charged ions are commonly observed in the nebulae and their study provides astronomers with powerful diagnostics of the physical conditions.

The neutral state of an atom is referred to by the Roman numeral I. Thus neutral hydrogen and carbon are labeled as HI and CI. Each time an electron is removed, the numeral is increased by one. Thus when hydrogen or carbon lose 1 electron, the resulting ion is labeled as HII or CII, respectively. Hydrogen only has one electron to lose. But carbon can lose up to six. Carbon that has lost three electrons is called CIV. A fully ionized carbon nucleus that has lost all six of its electrons is designated CVII. A gas consisting of a mixture of ions and electrons is called a *plasma*. Most plasmas are electrically neutral even though the constituent particles are not.

Chapter 2

1 Light waves:

Electromagnetic waves have the property that the product of frequency and the wavelength in vacuum equals the speed of light. The visual portion of the spectrum has wavelengths ranging from about 0.4 to about 1.0 microns (4000–10 000 angstroms). The regions between 912 Å (the longest wavelength that can ionize hydrogen) and 4000 Å is the soft-UV (or near-UV) region; the portion of the spectrum between 1 and 2.4 microns is the near-infrared. Between 2.4 and about 20 microns, thermal emission from the atmosphere and telescope is very bright, hence this object across the sky requires motion in both axes. Furthermore, the required rotation rate varies as a function of time and location in the sky. Thus, the use of alt–az mounts in astronomy requires the use of computers for tracking.

11   Arecibo antenna:

The 300 meter (1000 ft) diameter Arecibo radio telescope on the island of Puerto Rico is the largest filled aperture radio telescope in the world. Although its dish can not be moved, instruments at the focus can be moved to track astronomical sources for about an hour. The largest fully steerable radio telescopes are about 100 meters (300 ft) in diameter and all use alt-az mounts.

12   Adaptive optics:

Turbulence in the atmosphere introduces random fluctuations in the refractive index of the air which blurs the images produced by a telescope. At most observatory sites, this process limits the angular resolution of the sharpest images to about 1 arcsecond. *Adaptive optics* (AO) is the technology used to compensate for the wave-front errors introduced by the atmospheric turbulence. The distortions introduced by turbulence can be measured using a bright reference star in the field, or an *artificial star* generated by illuminating a spot in the upper atmosphere with a powerful laser beam.

A flexible mirror in the telescope is used to introduce a distortion to the incoming light that is opposite to that produced by the atmosphere. Turbulent distortions change on time-scales ranging from one-tenth to about one-thousandth of a second. Thus, the distortion-field has to be measured, and the corrections applied to the flexible mirror on this time-scale. The flexible mirror has to be equipped with as many actuators as possible to obtain the best correction. Typical adaptive optics systems use from 10 to over a 1000 individual actuators to compensate image turbulence. Adaptive optics works best at long wavelengths in the near-infrared. The resulting images can approach the diffraction-limited performance of large (8 to 10 meter diameter) telescopes at wavelengths of around 2 microns. Ground based AO has produced images with resolution better than 0.1 arcseconds.

13   Charting the Universe:

The Sloan Digital Sky Survey (SDSS) is using a 2.5 meter telescope to survey much of the northern sky to unprecedented depth. The spectroscopic part of this project has already found quasars and galaxies whose light started its journey towards us when the Universe was only 10 percent of its current age. The SDSS is the most ambitious survey to date that can claim to be mapping the Universe.

Chapter 3

1   Carbon monoxide:

Carbon monoxide (CO) is the best available tracer of cold gas in molecular clouds. Rotation of CO produces a series of bright spectral lines in the millimeter and sub-millimeter spectral region. The lowest energy transition in the most common form of CO has a wavelength of 2.6 millimeters (a frequency of about 115 GHz). Other transitions occur at 1.3 mm (230 GHz) and 0.87 mm (245 GHz). In warm and dense molecular cloud cores, there is a series of lines spaced at regular intervals of 115 GHz up to a frequency of several terahertz (10^{12} Hz).

Carbon monoxide comes in several isotopomers since both carbon and oxygen have several stable isotopes. Carbon has two stable isotopes, ^{12}C and ^{13}C, with the former being about 60–90 times as abundant as the latter. Oxygen has three stable isotopes: ^{16}O, ^{18}O, ^{17}O in order of decreasing abundance. The most common form of CO is $^{12}C^{16}O$ which has an abundance of about 10^{-4} times that of hydrogen in typical molecular clouds. The emission from $^{12}C^{16}O$ tends to only probe the surface layers of clouds since the cloud becomes opaque in the lowest energy emission lines of this species.

The second most common form is ^{13}CO. The greater mass of ^{13}C (it has an extra neutron) results in a 5 percent frequency shift in the spectrum. The lowest lying ^{13}CO transition occurs at about 110 GHz (and 220 GHz and 330 GHz for the next higher transitions). Most clouds are transparent in the ^{13}CO lines. Thus, this isotopomer is well suited for the study of the deep interiors of clouds.

Although molecular hydrogen (H_2) is the most abundant species in molecular clouds, it is very difficult to observe. Its lowest energy transition is at a wavelength of 28 microns in the thermal infrared. Furthermore, collisional excitation of this state only occurs in gas with a temperature of more than 100K.

H_2 can be observed in absorption against background stars and galaxies in the ultraviolet part of the spectrum from space. However, this method can only be used to trace H_2 in translucent clouds since dust in most molecular clouds is too opaque at UV wavelengths. Thus, CO is the best available tracer of the distribution of H_2 in molecular clouds.

2 Doppler effect:
The frequency f (or wavelength λ) of a spectral line is shifted by an amount directly proportional to the velocity. The shift is given by $\Delta f / f = \Delta \lambda / \lambda = V/c$, where V is the relative velocity of the source and observer and c is the speed of light. If the source and observer approach each other, the shift is towards shorter (bluer) wavelengths or higher frequency. This is called a "blueshift." If the source and observer recede from each other, the shift is towards longer (redder) wavelengths or lower frequency. This is called a "redshift."

3 Mass loss from red giants:
Most stars drive stellar winds. The Sun loses about 2×10^{-14} solar masses per year (about 4×10^{19} grams) at a speed of about 500 km/s. The solar wind is accelerated by the thermal expansion of plasma heated to a temperature of several million kelvin by magnetic activity and turbulence near the Sun's surface.

Red giant stars can lose more than 10^{-5} solar masses per year in low-velocity (10–100 km/s) winds driven by the pressure of star light acting on dust grains condensing in the red giant's atmosphere.

Luminous blue-supergiants also drive powerful winds with large mass-loss rates and high ($\approx 10^3$ km/s) speeds. These winds are accelerated by ultraviolet light pushing on a variety of common atoms and ions which have strong absorption lines in the UV portion of the spectrum.

4 Peak of black-body radiation:

The radiation from stars is roughly similar to the "cavity" radiation emitted by small holes in hot bodies: the so-called "black-body," or "thermal" radiation described by the Planck function. The peak of black-body emission shifts to shorter wavelengths according to the formula $\lambda_{peak}(\text{cm}) \approx 0.28/T\,(\text{K})$. Thus, the spectrum of a 1K body peaks at a wavelength of about 0.28 cm (2.8 mm), the spectrum of a 100K body peaks at about 0.028 mm (28 microns), and the spectrum of a 10 000K body peaks at about 0.000 28 mm (0.28 microns, or 2800 Å). On the low frequency (long wavelength) side of the peak, the spectrum increases in brightness proportionally to the square of the frequency. On the high frequency (short wavelength) side of the peak, the spectrum decreases exponentially with increasing frequency.

Fast electrons with speeds close to the speed of light spiral around magnetic fields. They produce a distinctly different type of radiation which increases in bright-ness towards *lower* frequencies: a trend opposite to the low-frequency behavior of black-body radiation. Therefore, this type of emission is called "non-thermal" radia-tion. Because this type of radiation can be emitted by particle accelerators known as "synchrotrons," it is also called "synchrotron radiation."

5 Pressure:

Heat is the result of random motions of atoms and molecules. The higher the temper-ature, the greater the motion. The random motions of particles results in collisions which can transfer momentum. Thus, the gas exerts pressure on its surroundings. In the absence of an inward force to confine the gas, it will expand in response to its internal pressure. The pressure is proportional to the gas particle volume density n and the temperature T and is given by $P = nkT = \rho V_s^2$, where ρ is the mass den-sity in grams per cubic centimeter. This pressure is proportional to the square of the mean speed of the constituent particles and the sound-speed squared (see next footnote).

6 Sound speed and particle motion in a gas:

The mean speed of particles in a gas, V, is related to the temperature T, and the masses, m, of the constituents (atoms or molecules) by $V^2 = 3kT/m$, where $k = 1.38 \times 10^{-16}$ (in c.g.s units) is known as Boltzmann's constant. This mean speed is proportional to the "sound speed" of the gas which is a measure of the propagation speed of sound waves. The "isothermal" sound speed is given by $V_s^2 = kT/m$.

For pure hydrogen atoms, $m_H = 1.67 \times 10^{-24}$ grams. Thus, at $T = 10$K, the mean particle speed is $V = 500$ m/s (≈ 0.5 km/s). At $T = 10\,000$K, $V \approx 16$ km/s. However, the mean mass for a cosmic abundance of hydrogen and helium, is 1.4 times greater than m_H so the mean particle speed in typical interstellar gas is about 20 percent lower.

In a photoionized HII region (nebula) with $T = 10\,000$K, nearly half of the par-ticles are free electrons. Thus, the mean molecular weight is around 0.7 (using a mean molecular weight of 1.4 for the neutral gas to account for the cosmic helium abundance) and the sound speed is around 11 km/s.

7   Shock waves:

Shock waves are formed when gas streams collide with velocities larger than the sound speed. As the flow enters the shock, the pressure and density increase abruptly at a sharp discontinuity called a *shock front*. As gas enters the front, the organized motion of its constituent particles become randomized by collisions with previous particles. Thus, the temperature of the gas also increases abruptly. The increase in temperature is proportional to the square of the velocity with which the gas enters the shock.

Interstellar gas entering a shock with a typical speed of several hundreds of kilometer per second can reach temperatures ranging from several hundred thousand to over a million degrees, sufficiently hot to completely destroy all molecules, fully ionize hydrogen, and to elevate most elements to high degrees of ionization. The subsequent recombination of electrons and ions produces a rich spectrum of emission lines and results in the cooling of the plasma.

Lower velocities generate less pressure and lower post-shock temperatures. If magnetic fields are present, they cushion the gas and limit post-shock heating and compression. Thus, magnetized shock waves tend to have cooler post-shock regions. Some molecules can even survive shocks with speeds less than 30 km/s.

As post-shock gas cools, its density increases, keeping the pressure approximately constant. For an interstellar mixture of elements, and a shock velocity of 300–500 km/s, post-shock gas can reach temperatures of about a million degrees. Recombinations will produce some X-ray and lots of UV radiation which cools the gas very rapidly to under 10 000K. The fully ionized portions of shocks emit bright lines of hydrogen such as Hα and lines from highly ionized species such as CIV, OIII, and SIII. Most shocks have a relatively extended tail of mostly neutral hydrogen where low-ionization species such as SII, and OII can produce bright radiation.

8   Cosmic rays:

High energy particles, or *cosmic rays*, are produced by magnetized shock waves, solar flares, neutron stars, and accretion disks around black holes. Cosmic rays consist of electrons, protons, some neutrons, and heavier atomic nuclei. Cosmic ray energies of cosmic rays are usually described in units of energy called *electron volts*, or eV for short. (1 eV corresponds to an energy of 1.6×10^{-12} ergs). A visual wavelength photon of light has an energy of about 2 eV. It takes photons with energies of at least 13.6 eV, or a wavelength of less than 912 angstroms, to ionize hydrogen atoms from the ground state. Most cosmic rays, especially the light electrons, move with speeds very close to the speed of light.

"Low energy" cosmic rays have energies of the order a million electron volts (MeV for short). But the spectrum of cosmic ray energies extends up to at least 10^{21} eV. Because most cosmic rays are electrically charged (e.g. they are bare electrons, protons, and fully ionized atomic nuclei), they gyrate in magnetic fields. As a result of their spiral motion, they radiate energy as low-frequency radio waves. Cosmic-ray electrons are responsible for the production of bright non-thermal radio emission from the Galaxy, from supernova remnants, pulsars, and jets powered by black hole accretion disks.

The collision of cosmic rays with gas and dust in the interstellar medium results in the deposition of energy which heats the gas. Cosmic ray collisions are the dominant heating mechanism in the interiors of molecular clouds where UV radiation is excluded. These collisions are also responsible for a small population of ions and electrons in the deep cloud interiors (where the fractional abundance of electrons is about 10^{-7} to 10^{-8}). This residual ionization keeps cosmic magnetic fields well-anchored to the interstellar gas.

Chapter 4

1   Ambipolar diffusion:
Charged particles are strongly coupled to magnetic fields. They are attracted to magnetic fields by the electromagnetic force just like the Earth is attracted to the Sun by gravity. As the Earth orbits the Sun, so charged particles orbit magnetic fields. But neutral particles feel no influence from magnetic fields.

In weakly ionized gas, the neutrals can "slip through" the charged particles anchored to magnetic fields. The slippage is called "ambipolar diffusion." The neutrals and the charged particles do occasionally collide, resulting in friction. More charges and/or higher density implies more friction and better coupling between charged particles and neutrals.

2   Angular momentum:
Angular momentum is a measure of the amount of rotational or orbital motion. The angular momentum, L, is the product of the mass, m, the distance, R, between the object and the point about which the angular momentum is to be calculated, and the component of the velocity, V_r, orthogonal to the distance r. Thus, the angular momentum about the point in question is given by $L = mV_r R$.

3   Conservation of angular momentum:
In the absence of torques or forces, the angular momentum of a body remains constant. Thus, if a spinning object such as the Earth experiences no friction, and is not acted on by any force to slow its spin, it keeps spinning for ever. If an isolated object contracts under the force of its own gravity, its spin rate increases so that the product RV_{spin} is a constant.

4   Hydrostatic core:
As a cloud core collapses, matter accumulates at the center. As the pressure (product of density and temperature) increases in the forming core, the inward motion of the collapse stops. The core enters a quasi-static configuration in which the internal pressure is in balance with the inward pull of gravity. More precisely, the rate at which pressure decreases with increasing distance from the center of the cloud is in balance with the gravitational attraction of the mass interior to a given location.

In this configuration, the core is said to be in "hydrostatic equilibrium." As the core surface radiates, the interior evolves towards an ever denser configuration. This slow contraction stops only when thermonuclear reactions ignite.

5   H-R diagram:

A plot of stellar luminosity (usually increasing towards the top on the vertical axis) versus stellar surface temperature (usually increasing towards the left on the horizontal axis); see Figure A1. About 90 percent of the known stars in the sky fall on an approximately diagonal line called the "main sequence." The main sequence of stars extends from the low-mass, type M "red dwarf" stars at temperatures of under 3000K and luminosities of under 10^{-3} times the luminosity of the Sun (1 solar luminosity corresponds to the rate at which the Sun gives off luminous energy, about 4×10^{33} ergs/s) to the high-mass, blue, type O stars with temperatures of more than 40 000K and luminosities about 10^6 times the Sun.

In between the most massive (about 100 solar mass) O stars and the least massive (0.08 solar mass) M dwarf stars, there is a monotonic progression of stars with decreasing luminosity and temperature. This sequence of normal stellar types is given the letter designations O, B, A, F, G, K, and M.

About ten percent of known normal stars lie off the main sequence and will be discussed later. Many more stars populate the low-luminosity region of the diagram than the high luminosity portion. Thus, there may be many more yet-to-be-discovered underluminous objects, including white dwarf, red dwarf, and brown dwarf stars.

6   Birth line:

Protostars start as very cool, low luminosity objects located in the lower right of the H-R diagram. They form a first hydrostatic core when cold molecular gas becomes opaque to its own radiation, at which point the core starts to heat. As it heats and compresses, the core dissociates all of its constituent molecules. A second phase of rapid core collapse occurs following dissociation. This collapse slows when hydrogen becomes fully ionized, and enters a second hydrostatic core phase. Slow contraction continues until the central temperature reaches a few million degrees at which point the rare gas deuterium (heavy hydrogen consisting of a neutron and a proton – its abundance is about 10^{-5} times that of normal hydrogen) starts to fuse into helium.

Deuterium fusion stalls further contraction until most of the protostar's deuterium is consumed on a time-scale of about 10^5 years. During this stage, low-mass protostars have higher luminosity and lower temperatures than main-sequence stars with the same mass. Stars in this stage are located on the upper right of the H-R diagram close to the red-giant branch. This locus of point is called the *birth line* since stars first become visible at visual and near-infrared wavelengths in this portion of the H-R diagram. The birth line has also been called the "deuterium main sequence."

7   Hayashi track:

Young stars evolve from the birth line to the main sequence; they migrate on the H-R diagram along trajectories known as *Hayashi tracks* and *Henyey tracks*.

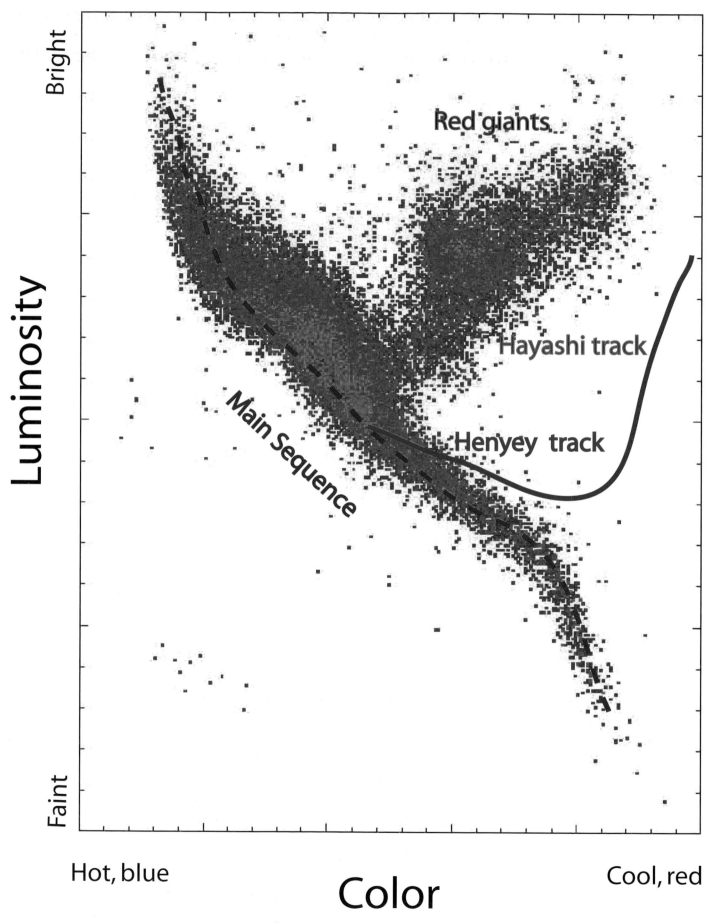

Figure A.1. A Hertzsprung – Russell diagram relates the temperature (horizontal axis) of stars to their luminosity (vertical axis). Temperature increases to the left, luminosity towards the top. (Hipparcos).

As deuterium is exhausted in the protostellar core by thermonuclear fusion into helium, the core contracts. The outer layers of the star respond by shrinking at approximately constant temperature. Thus the stellar luminosity declines and the star moves down along an approximately vertical line in the H-R diagram. During this phase, most of the energy generated in the core is transported to the stellar surface by convection. Thus, this is called the "convective" part or the Hayashi track.

As the stellar core heats to more than about 10^7K, all constituents become fully ionized and relatively more transparent than cooler, partially ionized gas located in the outer layers of the star. Energy can flow outward by radiation. Therefore convection in the core stops and the thermonuclear fusion of light hydrogen starts. As the size of the radiative core grows, the cooler and more opaque outer layers of the star continue to transport energy from the top of the radiative core to the stellar surface by convection, and the stellar surface warms at approximately constant luminosity. Thus, the star evolves horizontally across the H-R diagram until it reaches the main sequence. This is called the "radiative" part or the Henyey track.

Solar mass stars require about 15 to 20 million years to move from the birth line to the main sequence. Lower-mass stars take longer. A 0.1 solar mass star takes several hundred million years to make this transition while a 5 solar mass stars takes less than 1 million years.

Chapter 5

1   Red giant stars:

Stars located in the upper right of the H-R diagram (cool and over-luminous compared to the main sequence) are known as *red giant* stars. As core hydrogen is consumed by fusion into helium, the core shrinks, the stellar surface swells, and the star evolves horizontally on the H-R diagram towards lower surface temperature. Such "post main sequence" stars can once again become fully convective and increase in luminosity at about constant temperature. The trajectory of evolving low-mass stars on the H-R diagram roughly retraces the Hayashi tracks followed by forming stars.

2   End-states of stars:

Low-mass stars evolve off the main sequence towards the upper right of the H-R diagram: they become large and luminous, but very cool "red-giant" stars. The red giant stage is terminated by the formation of a "white dwarf" star from the stellar core and the ejection of the star's outer layer as a *planetary nebula*.

Stars with masses between about 5 to 20 times the Sun evolve off the main sequence by cooling at roughly constant (but slightly increasing) luminosity to become "red supergiant" stars. The cores of stars with initial masses of more than eight Suns eventually collapse to form rapidly rotating neutron stars. The sudden release of gravitational potential energy causes the outer parts of the star to explode as a hydrogen-rich "Type II" supernova.

Stars with initial masses above about 40 Suns never make it to the "red supergiant" phase: the pressure of their own light would blow off their outer layers. They evolve into short-lived and very rare "blue supergiant" stars, "luminous blue variable" stars, and peculiar "Wolf-Rayet" stars. These very massive stars tend to lose most of their outer layers and mass before their massive cores collapse into black holes. The release of energy powers a "Type Ib" supernova explosion deficient in hydrogen. Thus, the most massive stars produce black hole remnants.

3   Mass transfer and stellar variability:

When a member of a close binary evolves off the main sequence onto the red-giant branch, the star swells, and can, under some conditions, transfer some of its mass onto the companion star. Mass can also be transferred to the companion in the form of a stellar wind. Mass transfer is responsible for many types of of stellar variability.

In some systems, the companion is a compact and dense object such as a white dwarf, neutron star, or black hole (which implies that the companion was initially the more massive star in the system since it must have evolved faster). Matter from the swelling red giant will spill onto an accretion disk that forms around the companion. Accumulation of hydrogen on a white dwarf can lead to erratic variations in light output (cataclysmic variables), and occasional thermonuclear explosions on the star's surface as hydrogen is fused into helium (nova outbursts).

Mass transfer onto neutron stars is thought to be responsible for X-ray binaries and X-ray pulsars. Accretion onto black holes produce "micro-quasars." Accretion onto collapsed objects tends to be associated with the ejection of high velocity bipolar jets.

Continued accretion onto a white dwarf companion can drive it over the 1.4 solar mass stability limit. At just above 1.4 solar masses "degenerate" electrons can no longer support the star. It collapses, heats, and detonates in a thermonuclear explosion which rips apart the object in a "Type Ia" supernova explosion. Sustained accretion onto neutron stars may also push them over their maximum masses of about five solar masses, leading to the formation of a black hole.

4   Three-way encounters:

Two stars approaching along hyperbolic orbits cannot form a bound binary without dissipating some kinetic energy of motion. The presence of a third body is required to remove orbital energy. The third body can carry away some of the gravitational binding energy that has to be liberated to form a binary in a three-body encounter. Usually the two most massive stars form the binary, and the least massive star is accelerated and ejected by the interaction.

Chapter 6

1   Absorption lines:

Cooler gases located just above the stellar *photosphere* (the layer where most of a star's light is emitted into space) produce dark absorption lines in the spectrum of the star at the specific wavelengths where each type of atom has a transition which can absorb or scatter light. Thus the pattern of lines in a spectrum can be used to infer the chemical composition, ionization state, temperature, state of motion, and many other physical properties of the region producing the lines. Gases seen against a cooler or darker background radiate most of their energy by producing a set of bright emission lines.

2 Bipolar outflows:

When an object launches a pair of gas streams preferentially into two opposing directions, rather than uniformly into all directions, the flow is said to be *bipolar*.

If the light from a star is blocked in the equatorial direction, but can escape along both poles to illuminate a nebula, the result will be a *bipolar* nebula.

3 Destruction of clouds:

When a cluster of stars forms in a higher density and more massive cloud core, a greater fraction of the available mass may be converted into stars. The gas may be more tightly bound to the region, and outflows may have a more difficult time breaking out and disrupting the remaining gas. Thus, it may be more difficult to disrupt cluster-forming clouds. The final (total) mass of all stars formed in a cluster divided by the initial mass of the cloud, is called the *star formation efficiency*. The efficiency of star formation in a star forming region may be regulated by the combined impacts of outflows, UV radiation, stellar winds, and supernovae.

4 Coherent radiation:

Some processes (collisions, intense infrared radiation) excite more atoms or molecules into the upper state of a transition than are found in the lower state, leading to a peculiar condition known as a *population inversion*. A population inversion leads to the amplification of the intensity of radiation propagating through the gas. Radiation stimulates all atoms in the upper state to decay at once, resulting in extremely bright emission. This is the operating principle behind a *laser* (an acronym for Light Amplification by the Stimulated Emission of Radiation) and a *maser* (an acronym for Microwave Amplification by the Stimulated Emission of Radiation).

Several common interstellar molecules such as OH, H_2O, and SiO can develop inversions in hot, high density environments where there is an abundant amount of infrared light. These molecules produce very bright emission lines at radio wavelengths (18 cm for OH, 1.3 cm for H_2O, and 3.5 and 7 mm for SiO). The high intensity of these naturally occurring *masers* enables the use of very long baseline interferometry to measure the structures, locations, and velocities of the radiating gas elements with a resolution of about 0.0001 arcseconds.

5 Coupling of magnetic fields and gas:

The electron (and ion) fraction in the atomic phase of the interstellar medium ranges from about 10^{-4} to 10^{-6}. Low ionization potential elements such as sodium, potassium, and carbon can easily lose an electron when lit by soft UV. But, UV is excluded from the dusty cloud interiors. Deep inside molecular clouds, energetic cosmic rays produce an electron fraction (n_e/n_H) of order 10^{-7} to 10^{-8}, more than sufficient to strongly couple magnetic fields to the gas. Only in very opaque regions such as the densest cloud cores and the mid-planes of circumstellar disks does the ionization fraction drop to the point that magnetic fields decouple from the gas.

6   Dipole field:

Protostars have strong magnetic fields. These fields are initially dragged into the protostar during its collapse phase. The differential (nearly Keplerian) rotation of accretion disks tends to amplify this field until the pressure it exerts on the gas becomes comparable to the thermal pressure. Forming protostars are fully convective. Because most of their matter spirals in through the accretion disk, they also tend to spin fast. The combination of fast spin and convection is thought to result in further amplification of the magnetic field in the star by a process known as *dynamo action*.

The dominant component of the field is thought to be a dipole, a configuration which resembles a torus. The outer parts of the stellar magnetosphere threads the disk. The pressure exerted by a dipolar magnetic field tends to disrupt the inner 0.01 to 0.1 AU portion of the accretion disk. In the simplest models, the field strength increases with increasing stellar spin. Thus, the stronger the field, the farther out into the disk it can reach.

7   Magnetospheric accretion and disk winds:

Gas at the inner edge of the disk and the outer edge of the stellar magnetosphere is funneled to high latitudes by the field lines and thought to form giant bridges that arch onto the star. Magnetized stars can only accrete through these *funnel flows*.

The magnetic field tends to rotate with the star like a rigid body. Thus, the spin velocity of the field lines increases with increasing distance from the star. If the magnetic field spins faster than the accretion disk in the region where the disk is disrupted, the velocity difference acts to slow the stellar rotation. A slower stellar spin causes the field to weaken. The magnetosphere shrinks, and the disk inner edge can spiral closer to the star.

On the other hand, if the field is moving slower than the Kepler speed in the interaction zone, the rotation of the disk tends to increase the stellar spin rate, resulting in a stronger field that drives the disk inner edge farther from the star. Thus dynamo action and interaction of the stellar magnetic field with the disk may regulate the spin of forming stars.

These magnetic fields are also thought to launch the jets and winds that drive Herbig-Haro objects. Field line bundles that rotate faster than the local orbit speed tend to sweep up and accelerate material. This gas then tries to move away from the star to a larger radius orbit; it drags the magnetic field with it. If the gas pressure and inertia exceeds the magnetic field pressure, the interaction can accelerate the gas to escape speed and create open field lines. Gas can be continuously accelerated by these open field lines into the oppositely directed lobes of a bipolar outflow that emerges along the rotation axis of the star.

Chapter 7

1   Hα line:

When the electron in hydrogen jumps from the third excited state (energy level) in hydrogen (n = 3) to the second energy level (n = 2), it emits light at a wavelength of 6563 angstroms. This transition in atomic hydrogen is known as the Hα line.

2   Keplerian rotation:
A particle moving in a stable circular orbit with radius R in the gravitational field of a star of mass M moves with a velocity $V = (GM/R)^{1/2}$, where $G = 6.67 \times 10^{-8}$ in c.g.s. units. Thus, the orbital speed declines as the inverse-square-root of the distance from the star.

3   Angular momentum transfer in disks:
Conservation of angular momentum in a disk implies that as matter spirals towards the central star, its angular momentum must decrease. (The Kepler speed increases as the square root while angular momentum conservation implies that the speed should increase linearly with decreasing distance.) Magnetic fields may extract angular momentum from the inner disk and transfer it either to the outer disk, or to a wind. Turbulence, spiral waves, dense clumps, protoplanets, or companion stars may also extract angular momentum and enable accretion to occur.

4   Spectral types of stars:
The spectral designation of stars ranges from hot O to cool M. In order of decreasing temperature, the spectral sequence consists of types O, B, A, F, G, K, and M. Cooler brown dwarfs have been given the designations L and T.

5   Convective cores and radiative envelopes:
M dwarf stars are fully convective. Higher mass main-sequence stars have radiative cores and convective envelopes. The relative size of the radiative core increases and the thickness of the convective shell decreases with stellar mass. Moderate mass stars (2 to 6 solar masses) have very thin, convective shells. Surface convection disappears altogether in massive stars. However, O and B stars develop convection zones within their cores.

Chapter 8

1   Random motions of stars:
The random motions of stars tend to be similar to the escape speed from the parent cloud. In low-mass cloud cores, random motions are typically about 1–2 km/s, similar to the Doppler widths of molecular emission lines from cold cloud cores.

2   Star formation efficiency:
The *star formation efficiency* is defined as the ratio of the total mass of stars formed divided by the total initial mass of a star-forming cloud. To form a gravitationally bound cluster, the star formation efficiency has to be be at least 30 percent because of the so-called *Virial theorem*. According to the Virial theorem, the average kinetic energy of a swarm of objects bound to each other by gravity will be one-half of their gravitational potential energy when the system is in *equilibrium*. Consider a swarm of cloud cores in a molecular cloud, some of which will form a star. According to

the Virial theorem, the typical random velocity of each core (or resulting young star) as it moves around in the gravity well of the entire cloud will be $(1/2)^{1/2}$ of the gravitational escape speed from the cloud.

If, once star formation is complete, as much mass (or more) than is contained in all the young stars is promptly removed, this average speed of the stars will be equal to (or greater than) the escape speed from mutual gravity of the cluster and most of the stars will disperse. This can be seen from the formula for the gravitational escape speed. A test particle (a star) moving about the gravitational field of a cluster of mass M, here idealized as a point mass located a distance d from the star, is given by $v_{escape} = (2GM/d)^{1/2}$, where G is Newton's gravitational constant. The orbital speed of the star about the cluster is on average $2^{1/2}$ smaller than the escape speed. If half (or more) of the cluster's mass were suddenly removed, the escape speed from the cluster would decrease by $2^{1/2}$ and the test particle's previous velocity would now be just equal to (or greater than) the new escape speed.

If gas is removed on a time-scale much longer than the time it takes a typical star to cross the cluster, the cluster can remain bound as an open cluster, even for very low star-formation efficiency.

However, for many forming clusters, the gas is removed on a time-scale that is short compared with the cluster-crossing time. (Ionized gas can expand at about 10 km/s, the sound speed in an H II region, while the typical escape speed from an Orion-like cluster is only a few kilometers per second.)

As the parent cloud is dispersed, all clusters expand. If the total mass of new stars is less than the total mass of gas removed at the end of star formation, then most of the stars will form an expanding association. On the other hand, if the total mass of new stars is greater than the mass of gas removed, most of the stars will form a bound cluster. The smaller the ratio of dispersed mass of gas divided by the total mass of stars, the less expansion the cluster experiences.

3 Supernovae:

Supernovae are the 10^{51} erg explosive deaths of stars. They come in two general varieties: Type I and Type II. The former lack hydrogen in their spectra while the latter have it. Type Ib and Type II supernovae are powered by the release of gravitational potential energy; the Type Ia explosions are thermonuclear detonations of white dwarfs.

The iron cores of evolved stars with initial masses of more than about eight but less than about 50 solar masses explode as Type II supernovae which can leave behind neutron star remnants. More massive stars often lose their hydrogen-rich envelopes before exploding. Thus, these stars lack hydrogen in their spectra. These most massive stars explode as the so-called Type Ib supernovae which leave behind black hole remnants. (See also Chapter 14 note 7.)

4 Late-type spirals and irregular galaxies:

Spiral galaxies with very small bulges and irregular spiral arms are called "late-type" spirals.

Some galaxies are completely disorganized with no clear spiral patterns in the distributions of stars and gas. These objects are called "irregular galaxies." See chapter 15 for more discussion.

5   Dwarf elliptical galaxies:
Small galaxies consisting of only old stars, usually distributed in a smooth spherical or elliptical configuration, are called "dwarf elliptical" systems. These galaxies have very little interstellar gas and dust and tend not to support any ongoing star formation. See chapter 15 for more discussion.

6   The spheroidal bulge and halo of our Galaxy:
Most spiral galaxies contain flat disks, and central bulges. The bulge and diffuse halos consist of old stars and tend to be round or elliptical. Bulges often resemble elliptical galaxies. See chapter 15 for more discussion.

7   Planetary nebulae:
Planetary nebulae are formed when the outer layers of a red giant are expelled and ionized from the inside by the stellar core that is evolving towards becoming a white dwarf.

Chapter 9

1   Distance between stars:
Near the Sun, there is one star in about every 30 cubic light years. But in the center of the Trapezium cluster in the Orion Nebula, the density of young stars is more than 3000 stars per cubic light year; 100 000 times greater than the density of stars near the Sun. However, there are only about 40 stars in this concentration. Young stars are packed so close in the Orion Nebula's core that their separation is only a few thousand times the distance between the Earth and the Sun.

If the spacing between stars is one hundred times smaller than near the Sun, the number of stars in a given volume of space increases by a factor of a *million*! As the stellar spacing in a cluster shrinks, the probability of chance encounters increases dramatically.

2   Orion:
The constellation of Orion consists of four bright stars in a large rectangle. *Orion's Belt* consist of three distinctive bright stars in a row located in the middle of the rectangle. *Orion's Sword* consists of a north-south string of faint stars about 5 degrees below Orion's Belt.

3   Another ionized nebula born from Orion A:
This ionized nebula is known as NGC 1977. Another ionized nebula, known as M 43, lies just north-east of the Orion Nebula.

4   Another ionized nebula born from Orion B:
This region, located near the eastern Belt star, Zeta Orionis, is called NGC 2024.

5   The Orion 1b and 1d sub-groups:
The Orion 1c sub-group contains the clusters NGC 1981 about a degree north of the Orion Nebula and the stars surrounding Iota Orionis. These stars are located about 50 to 100 light years in front of the Orion Nebula. The Sigma Orionis group is also considered to be part of Orion 1c although it is somewhat younger. Theta Orionis in the Orion Nebula and NGC 1977 are embedded within the Orion A cloud and are thus younger. These two stellar groupings are considered to be part of Orion's youngest sub-group, Orion 1d.

Chapter 10

1   Early theories of Solar System formation:
The first theory of the formation of the Solar System based on Newton's ideas of gravity and mechanics was proposed over 200 years ago by Kant and Laplace. They hypothesized that the Solar System formed from a primordial cloud which had some spin. Collapse formed a disk from which the planets would emerge as small particles coagulated into larger ones. The Sun formed from gas in the center of the disk. This early theory is remarkably similar to the modern "standard model" of planetary system formation. Kant and Laplace would be amazed by the images returned by HST which show that disks are visible around many recently formed stars.

2   Capturing planetesimals by proto-Jupiter's atmosphere:
Gas giant protoplanets are large. In the gravitational instability scenario, the initial condensation may be as large as an AU in diameter. Even in the standard core-accretion model in which the hydrogen-rich atmosphere is accreted only after the formation of a large rocky planet, the initial diameter of a giant planet is very large. Regular moons may have formed *in situ* from a circumplanetary disk fed by the circumstellar disk. Alternatively, they may be planetesimals captured and trapped by the gas giant's extended early atmosphere.

Chapter 11

1   Chondrules:
Chondrules are spherules of rocky material which was melted in a zero-gravity environment and then re-solidified and incorporated into primitive Solar System bodies. Their sizes range from hundreds of microns to many millimeters in diameter. There are several theories for their origin, but no consensus has yet emerged about which model is correct.

2   Extinct radioactivity:
Inclusions in meteorites that contain anomalous isotopic abundances provide conflicting evidence about the type of environment in which the Solar System formed. Some meteorite samples indicate contamination of the parent clouds and proto-Solar nebula by debris from a red giant star. Other evidence points to pollution by supernovae. Some samples contain the daughter products of short-lived radioactivities

with half-lives of less than a million years. The parent species were probably synthesized in the core of a massive star and expelled by a supernova explosion no more than a few million years prior to being locked into the meteoritic inclusions in which they decayed to their daughter species. This type of evidence indicates that the Solar System was likely born in an OB association in close proximity to massive stars. Thus the forming Solar System and its evolving disk may have been influenced by concurrent supernovae. This point will be discussed further in Chapter 12.

3   Heating of planetesimals:
Meteorites provide evidence that planetesimals larger than about 100 km were internally heated by radioactive decay. The intensity of early radioactivity in the young Solar System provided indirect evidence for the injection of short-lived isotopes produced in massive stars a few million years before the formation of the Solar System.

Chapter 12

1   Photo-ionization:
Ultraviolet light can strip the electrons out of atoms, leaving behind positively charged ions. This process is called *ionization*. Hydrogen, the most abundant element in the Universe, can be ionized by hard-UV radiation with a wavelength less than 912 angstroms. Longer wavelength photons do not have sufficiently large energy to ionized hydrogen from its ground state.

The photon energy (in ergs) is given by $E = h\nu$, where $h = 6.626 \times 10^{-27}$ is Planck's constant in centimeters–grams–seconds (c.g.s.) units and ν is the photon frequency. The photon frequency is related to its wavelength by $\nu = c/\lambda$, where $c = 2.998 \times 10^{10}$ cm/s is the speed of light and λ is the wavelength of the photon in centimeters. Photon energies are often given in terms of an electron volt (eV), where 1 eV corresponds to an energy of 1.602×10^{-12} ergs. A 912 angstrom photon has an energy of 13.6 eV. The energy required to ionize hydrogen from its ground state is 13.6 eV. Thus, more energetic photons ionize hydrogen out of its ground state while less energetic photons do not. Carbon, the fourth most abundant element in the Universe, can be ionized by photons with energies larger than 10.2 eV. Thus, soft-UV can ionize carbon.

2   Proplyds:
Figure A2 illustrates the processes that shape proplyds.

3   Viscosity:
The sources of viscosity in protoplanetary disks are poorly understood. Magnetic fields, spiral density waves, and turbulence are likely possibilities.

4   Decay products:
Inclusions in primative meteorites often contain the decay products of short-lived radioactive species such as ^{26}Mg, the stable isotope produced by the decay of ^{26}Al (half-life of roughly one million years).

Figure A.2. Complex physical processes shape the structure and evolution of proplyds. (J. Bally/STScI).

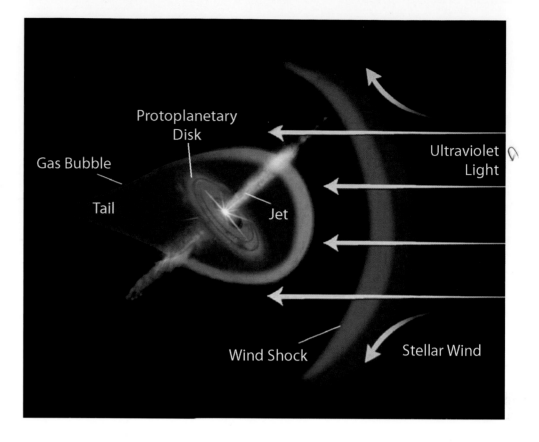

Some short-lived radioactive species, such as ^{27}Al, can be produced by energetic protons generated in stellar flares impacting stable isotopes such as ^{26}Mg. However, others, such as ^{60}Fe can only be produced in the neutron-rich environment of a supernova.

Thus, the detection of decay products of ^{60}Fe in meteoritic chondrules provides strong evidence that our forming Solar System was polluted by the radioactive debris of a nearby supernova.

5 Orbital migration:

When large planets form from a disk, they create gaps around their orbital radii. Thus, they become isolated from direct contact with the disk. However, the gravity of the planet can excite density waves in the disk which can transport angular momentum. Viscous evolution of the disk will transfer angular momentum outward, and result in the migration of the disk inward. Any planets and their gaps will also experience orbital migration. Inner planets tend to give-up angular momentum and will therefore be forced to drift towards the parent star; planets near the outer edge of the disk will tend to absorb angular momentum, and thus drift to orbital radii. The "hot gas giants" discussed in Chapter 13 were probably formed at large radii, migrated close to their parent stars, and were then parked there as the disk was dissipated. Planets formed during early phases of disk evolution may migrate so close to the parent star that they are accreted.

Chapter 13

1   First discoveries of extra-solar planets:
An excellent recent account of the search for extra-Solar planets is given in *Looking for Earths* by Alan Boss (1998, John Wiley and Sons). Other references to the original discovery papers include Mayor and Queloz (1995, Nature, 378, 355), Marcy and Butler (1996, ApJ 464, L147), Butler and Marcy (1996, ApJ 464, L153).

2   51 Pegasi:
The radial velocity curve of 51 Peg shows a sinusoidal radial velocity variation with an amplitude of about 60 meters per second at the orbital period of about 4 days.

3   Transits:
The transits of Mercury and Venus observed from different locations can be used to synchronize clocks and, combined with measurements of the elevation of the Sun and stars, to measure longitudes. The much more frequent transits and eclipses of the moons of Jupiter can be used for the same purpose. A major use of early observations of the transits of Venus was the determination of the Astronomical Unit (the Earth–Sun distance distance) from the parallax of the transiting planet observed from different sites.

4   HD 209458:
Though only a 1 percent effect, transiting hot Jupiters can be easily detected with precision photometry. Space-based photometry can detect 0.01 percent changes in light amplitude, and 0.001 percent changes with some care.
   David Latham of the Harvard-Smithsonian Center for Astrophysics in Cambridge, Massachusetts, inferred the presence of a gas giant planet in orbit around HD 209458 from analysis of radial velocity data and urged the STARE group to monitor this star for transits. The first transits were detected exactly at the expected times deduced from the radial velocity data.

5   JWST:
For further details, see http://ngst.gsfc.nasa.gov

6   TPF:
For further details, see http://planetquest.jpl.nasa.gov/TPF/tpf_index.html

Chapter 14

1   Thermonuclear burning of hydrogen:
Fusion of hydrogen requires four protons, two of which have to be converted into the slightly more massive neutrons. The rest mass energy of ^{4}He is 0.007 percent lower than the rest mass energies of four protons. The energy is liberated by the fusion of H into He and emerges partly as invisible neutrinos and partly as sunlight.
   Low-mass stars predominantly fuse hydrogen into helium by a process known as the proton-proton (P-P) chain. Stars more massive than about two solar masses utilize the catalytic carbon-nitrogen-oxygen (CNO) cycle.

In the H fusion, energetic collisions between protons result in the conversion of one proton into a neutron via inverse-β decay. In the quark model of nucleons, a proton contains two "up" quarks and one "down" quark and the neutron contains one "up" quark and two "down" quarks. Thus, the formation of a neutron from a proton involves the conversion of one "down" quark into an "up" quark. In this process, the electric charge of the proton is shed by the emission of a positron (an anti-electron), and the creation of a weakly interacting particle called a neutrino.

In the P-P chain, two protons (H) collide to form a deuterium (heavy hydrogen) nucleus consisting of a neutron and a proton. Then, a collision between H and deuterium forms ^{3}He which consists of two protons and one neutron. In the most probable branch of the P-P chain, two ^{3}He collide to form ^{4}He, a process that ejects two protons.

In the CNO cycle, a series of proton collisions with carbon, and nitrogen, and oxygen lead to the formation of ^{4}He and the recycling of the carbon. In the first step, a collision between H and ^{12}C forms unstable ^{13}N which decays by converting a proton into a neutron to form ^{13}C. Another proton collision forms ^{14}N. Then, a proton collision converts ^{14}N to ^{15}O which decays to ^{15}N. Finally, ^{15}N plus a proton collide to produce ^{12}C and ^{4}He. The net energy output is the same as for the P-P chain.

The CNO cycle can only operate when some carbon is present. In Chapter 16, we discuss the consequences of having no carbon: a situation that arises in the very first generation of stars in the Universe. See also note 8 to Chapter 1.

2   Triple-α process:

A helium nucleus is known to physicists as an *alpha* particle. (Energetic electrons are known as *beta* particles and energetic photons in the gamma-ray portion of the spectrum are known as *gamma* particles. This terminology dates back to an empirical classification scheme of the types of particle produced by radioactive decay which was discovered at the end of the nineteenth century.) Thus, in the triple alpha process, three helium nuclei (each containing two protons and two neutrons) come together and react to form a 12-carbon nucleus consisting of six protons and six neutrons.

3   Death of very low-mass stars:

Although isolated stars with less than about 0.8 solar masses have main-sequence lifetimes longer than the current age of the Universe, there are several mechanisms which can truncate their lives prematurely. A low-mass star in a short period binary system can be swallowed by its companion when it enters the red giant phase. For slightly larger orbital separations, mass transfer from the red giant can add enough mass to shorten the low-mass star's life. In galactic nuclei, low-mass stars can be ripped apart and swallowed by giant black holes. There are other extremely rare situations which may prematurely terminate a low-mass star's main-sequence life.

4   The solar wind:

The Sun's stellar wind, also called the *solar wind* blows with a speed of around 400 to 800 kilometers per second. The average mass-loss rate of the solar wind is about

2×10^{-14} solar masses per year. Studies of other, younger Sun-like stars indicate that the solar wind may have been stronger in the past. Even so, over its 4.5 billion year history, the Sun has lost a negligible fraction of the mass it was born with. The solar wind is thought to be powered by the thermal expansion of the hot corona: the solar wind is said to be a "thermally driven" wind.

In contrast, the massive O-type stars have mass-loss rates of the order of 10^{-8} to 10^{-5} solar masses per year and wind velocities higher than 1000 kilometers per second. The mass-loss rates increase with increasing luminosity of the star. These powerful O-star winds can therefore reduce a massive star's mass significantly over its lifetime. For example, a 30 solar mass O7 spectral type star typically loses mass at a rate of several times 10^{-7} solar masses per year. Over a 10 million year lifetime, such a star will lose several solar masses. The most massive stars may lose most of their mass by the time their cores exhaust hydrogen and the stars evolve off the main sequence.

Hot star winds are predominantly driven by the pressure of star-light coupling to strong UV absorption lines. Thus, these are said to be "line-driven" winds. In some very luminous and massive stars, the radiation pressure couples directly to electrons, resulting in a "continuum-driven" wind.

As stars become giants and supergiants, their mass-loss rates usually increase, sometimes quite dramatically to rates ranging from 10^{-5} to higher than 10^{-3} solar masses per year. Red giant atmospheres are sufficiently cold to allow the condensation of refractory dust grains. These winds are powered by radiation pressure acting on the dust grains.

5   Red and blue supergiant phases of massive stars:
High-mass stars evolve off the main sequence as they complete the conversion of H to He in their cores to become red supergiants. However, stars with initial masses greater than about 50 Suns are too luminous to become red supergiants; the radiation pressure associated with their immense luminosities would blow off their outer layers. Stars with greater mass become blue supergiants during their last evolutionary stages that precede their supernova explosions. The most massive stars occasionally suffer great eruptions during which they can lose many solar masses. Such mass loss can expose their helium, carbon, nitrogen, or oxygen enriched cores. The stellar cores exhibit peculiar spectra dominated by these products of nuclear burning.

6   Nuclear binding energy curve:
Iron is the most tightly bound element in the periodic table. Nuclear energy can be extracted either by fusion reactions that process lighter elements into heavier ones up to iron, or by fission reactions in which elements heavier than iron break apart to form lighter ones. Thus, iron is said to be the lowest point in the nuclear binding energy curve.

7   Supernovae:
There are two generic types of supernovae, called type I and type II, based on the absence or presence of strong absorption due to hydrogen in their spectra. A so-called core-collapse type II supernova being described here releases about 10^{51} ergs

of energy (10^{44} joules). Type II supernovae are the explosions of massive stars resulting from the collapse of their iron cores. The most massive stars, those above a mass of about 60 times the mass of the Sun, can lose so much mass during their main sequence evolution that by the time such stars explode as supernovae, they have completely expelled all hydrogen from their outer layers. Such massive supernovae do not contain any spectral lines due to hydrogen. Thus, they are classified as supernovae of Type Ib or Ic (depending on the heavy elements present in their spectra).

The Type Ia supernovae also lack any signs of hydrogen in their atmospheres. These explosions are thought to be produced by the thermonuclear detonation of white dwarf stars that have accreted enough matter from a closely orbiting companion so that their masses exceed 1.4 times the mass of the Sun. White dwarfs above this mass are not stable and they collapse under their own weight. The release of gravitational potential energy drives the thermonuclear burning of the helium/carbon mixture that tends to dominate the composition of such white dwarfs. The nuclear conflagration blows the collapsing white dwarf star completely apart. Thus, one can think of a Type Ia supernova as a thermonuclear explosion of the helium/carbon white dwarf that has been pushed over the critical mass of 1.4 solar masses. Type Ib/c and Type II supernovae are gravity bombs in which the formation of a neutron star or black hole releases gravitational potential energy. Both types of explosion release about the same amount of energy. However, the most massive Type Ib/c and Type II events may approach an energy scale of 10^{53} ergs.

8   Supernova shocks:
While the Herbig-Haro objects tend to emit in neutral and low-ionization tracers such as neutral oxygen and singly-ionized sulfur, supernova remnants are often bright in the lines of doubly-ionized oxygen and other highly ionized species. The gas behind a supernova shock can reach temperatures of millions to hundreds of millions of degrees and emit large amounts of X-ray radiation.

9   The lowest-mass supernova progenitors:
An eight solar mass star corresponds to spectral type B3. Lower mass stars, those with spectral types B4 or later, do not have sufficient mass to end their lives in supernova explosions.

10   Gould's Belt:
The Gould's Belt is defined by a band of bright blue B and A stars that surround the Sun's neighborhood. These relatively massive stars are short lived with ages less than about 100 million years old. Thus, they serve as tracers of relatively recent star formation in the solar vicinity. The Gould's Belt is dominated by the nearby young OB associations such as Sco-Cen and Orion with ages of 15 million years or less. It also includes some "fossil" OB associations such as the Alpha Persei cluster and the Cas-Tau groups which have lost all stars more massive than about seven or eight solar masses.

11　21 cm emission:

In the early 1940s, the Dutch astronomer Jan Oort (after whom the Oort cloud of comets is named) realized that a radio frequency transition could be used to study the structure of the Milky Way galaxy unimpeded by the obscuring effects of interstellar dust. The 21 centimeter wavelength (1.4 GHz or 1.4×10^9 Hz) radio spectral line of atomic hydrogen was discovered immediately following World War II. It has become the principal tracer of the distribution of cold neutral atomic hydrogen in the Milky Way. The 2.6 mm (115 GHz) line of carbon monoxide (CO) was discovered in 1970 and soon became the best tracer of the molecular gas in the Galaxy.

12　Baade's hole:

Walter Baade identified a region in the sky which contains less gas and dust than any other line of sight. This "hole" provides an exceptionally clear view of distant galaxies. Baade's Hole lies directly above the Alpha Persei cluster and the Cas-Tau group and provides evidence that an ancient superbubble erupted from this region and blew a large hole in the distribution of gas and dust in the Galactic plane (see Figure A3). There is a less obvious hole toward the southern sky directly below the Cas-Tau group. However, the Sun's current position places us about 30 parsecs from the middle of the Galactic plane. This makes the northern hemisphere hole much more obvious. Baade's Hole is also a well-defined hole in the all-sky distribution of atomic hydrogen; it was first noted to be the line of sight with the lowest column density of HI by Jay Lockman. Thus, it is also known as the "Lockman Hole." Baade's Hole should not be confused with Baade's Window, a particularly clear line of sight toward the bulge of our galaxy.

13　Fragmentation of super-rings:

As the shells swept up by superbubbles become more massive and slow down, their own self-gravity can come to play important roles in their development. As they age, self-gravity tends to fragment shells into gravitationally bound clouds. But, both small and very large portions of supershells are stable with respect to gravity. Small pieces do not contain enough mass for gravity to overcome internal motions. On the largest scales, the ring's expansion will usually be faster than the gravitational escape speed from the swept-up shell. However, gravity can become important on intermediate scales. Self-gravity dominates when the region enclosed by an imaginary sphere contains enough mass so that the gravitational escape speed from the sphere's surface is larger than the average value of internal motions.

Chapter 15

1　The largest galaxies:

A typical galaxy like the Milky Way is about 100 000 light years in diameter and contains about 10^{11} stars. While some dwarf elliptical galaxies have as few as 10^6 stars (not many more than typical globular clusters), most dwarfs have between 10^8

Figure A.3. A cluster of massive stars born in the Galactic plane forms a superbubble, which bursts out of the plane to form two holes above and below the plane. (J. Bally).

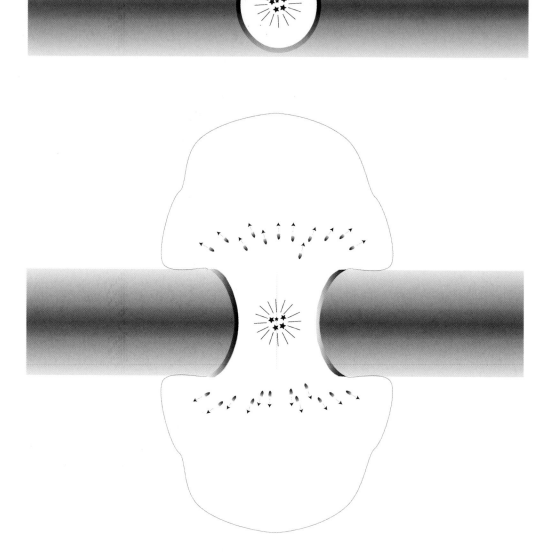

and 10^{10} stars. The giant ellipticals at the centers of rich galaxy clusters, the so-called cD (central-dominant) ellipticals, can have between 10^{13} to 10^{15} stars.

2   Our Galaxy from the outside:
Our Milky Way is an Sb/Sc spiral with a very small and weak bar and a bulge of stars that extends to a radius of about 3 kpc from the Galactic center. It may resemble the galaxy M83 shown in Figure 3.2 except that the Milky Way may have a larger bulge, smaller bar, and a more pronounced ring of molecular clouds peaking around 4–to 5 kpc from the Galactic center.

3   Dark matter and dark energy:
The Universe appears to consist of about 4 percent ordinary matter (see footnote in chapter 1), 25 percent dark matter, and 70 percent dark energy. Dark matter exerts gravitational force and can be mapped by the gravitational lensing effect. Evidence

comes from the motions of galaxies in galaxy clusters, the lumpiness of the cosmic microwave backgrounds, from the rate at which large-scale structures form under the influence of gravity in the cosmos, and from the flat rotation curves of galaxies.

The nature of dark matter remains unknown. However, recent theories in particle physics posit the existence of an entire family of weakly-interacting (with normal matter) "supersymmetric" particles which have many of the properties required of dark matter. Each particle that we currently know (such as the photon, electron, proton, etc.) has a supersymmetric partner with a spin that differs by a half-integer. For example the spin-1/2 electron's supersymmetric analog is called a "selectron" and the spin-1 photon has a half-integer spin partner called a "photino."

Dark energy is now suspected to exist because of an apparent acceleration in the expansion rate of the Universe. It appears to act on normal matter like a repulsive form of gravity produced by the vacuum.

The nature (and reality) of dark matter and dark energy provide the greatest challenges to modern physics and astrophysics.

4 Ram pressure:

When a fluid or gas moves past an obstacle with a velocity V, the pressure on the obstacle exerted by this bulk motion is given by the product of the density, ρ, and the velocity squared. This pressure, given by $P = \rho V^2$ is called the "ram pressure."

5 Collisions of stars:

The typical separation between stars in a galaxy is about a light year (one light year $\approx 10^{18}$ cm). And, if you look at a typical distant galaxy, the projected *surface density* of stars is about 100 stars per square light year. When two galaxies collide, the chance than any one star hits one in the incoming galaxy is about 1 in 10^{12}. So, typically one might expect at most a few violent star–star collisions in a galaxy merger.

6 The black hole in M106:

Ultra-high resolution radio interferometric observations have revealed the presence of a dense disk of gas orbiting the center of the galaxy M106 with velocities up to 1000 km/s. These high velocities provide the best direct evidence to date for the presence of a supermassive black hole in the center of a galaxy other than our own.

7 Active galactic nuclei:

Giant black holes appear to be common in the nuclei of galaxies. They range in mass from under 10^6 solar masses to about 10^{10} solar masses. The masses of central black holes appear to scale with the mass of the *bulge* (that portion of a galaxy that resembles an elliptical).

Quasars (an acronym for QUAsi StellAr Radio source) are thought to be nuclear black holes accreting matter from their environments. Such matter tends to form a thick torus surrounding the hole. The outer parts of such tori have diameters ranging from a fraction of a light year to hundreds of light years and consist of cold molecular gas. The tori get thinner, hotter, and denser with proximity to the black hole. Their inner regions can reach temperatures hot enough to emit X-rays. As matter spirals in through the disk, entwined magnetic fields drain angular momentum, launch powerful jets, and accelerate electrons and other charged particles to relativistic energies. These processes produce powerful emission throughout the spectrum. Quasars are the most luminous sources of electromagnetic energy in the Universe that produce relatively steady emission (gamma-ray bursters can outshine quasars for a few minutes). Most lie at cosmological distances ranging from a billion to 13 billion light years. The most luminous quasars can have luminosities in excess of 10^{14} times the luminosity of the Sun. They were more abundant in the early Universe when galaxy collisions, and consequent feeding of central black holes, was more common than today.

Accretion onto black holes is responsible for several types of Active Galactic Nuclei (AGN). When the disk is oriented so that we can see its bright inner parts, the object is seen as a quasar. Some quasars are radio-bright while others exhibit only faint radio emission. In most radio-loud quasars, the emission is produced by a highly collimated jet. These jets often have components which move at nearly the speed of light. When the jet-beam is aimed directly at us, the strength of the emission from the jet can be greatly enhanced (by so-called *Doppler boosting*). When the synchrotron emission from the jet overwhelms the light of the underlying accretion disk and host galaxy, it produces a nearly featureless spectrum and the object is called a *BL Lac* object after the prototype.

In some AGN, the radio jets inflate large cocoons of radio-bright plasma which are seen as a pair of radio-bright regions on either side of a galaxy (usually an elliptical). When the circumnuclear torus obscures the black hole and inner accretion disk, only the radio jet or lobes are visible. Such objects are called radio galaxies.

AGN less luminous than quasars are called Seyfert galaxies. They come in two varieties. Type I Seyferts have broad emission lines, probably produced by the inner accretion disk. Type II Seyferts have only narrow emission lines. It has been argued that the difference between Type I and Type II is an orientation effect; Type I Seyferts are seen face-on while Type IIs are edge-on so that the inner accretion disk responsible for the broad emission lines is obscured.

Chapter 16

1   Black-body spectrum:
    Black-body spectra (see Chapter 3 note 4) are generated when the radiation field is in thermodynamic equilibrium with the temperature of matter. A cavity in a hot body produces an excellent realization of an ideal blackbody radiation field. An opaque gas or plasma tends to eventually come to equilibrium with the radiation field that it produces. Thus, the cosmic microwave background is very well described by as a black-body radiation field.

2   The opaqueness of the cosmic microwave background:
    As discussed in Chapter 1, the cosmic microwave background (CMB) was produced when the hot plasma emerging from the Big Bang first recombined to form atoms of hydrogen. While the hydrogen is very transparent, the plasma of electrons and

protons that preceded the era of recombination is not. Free electrons in the plasma effectively scatter radiation at all wavelengths. Thus, we cannot see beyond the CMB with electromagnetic waves at any wavelength. However, in principle, much earlier phases of the young Universe can be probed directly by non-electromagnetic radiation produced at even earlier times. Neutrinos and gravity waves are two possible examples of such radiation.

3   Light exerts pressure:
As discussed in earlier chapters, radiation pressure limits the masses of the most luminous stars since the pressure of light emerging from the stellar interior exerts a force on electrons and ions which becomes comparable to the force exerted by gravity. Any increase in radiation pressure will tend to drive away the outer layers of the star, thereby decreasing its mass. Prior to recombination, radiation pressure in the cosmic plasma greatly dominates the force of gravity due to local density fluctuations and therefore tends to smooth out any nonuniformity. Dense clumps will tend to expand into surrounding lower density regions. Dark matter, on the other hand, does not interact with light, and can therefore form clumps on its own before recombination. The gravitational field of such clumps can then start to attract surrounding plasma. But, until after recombination, the resulting density enhancements in the plasma are kept smaller than a few parts in 10^5 by radiation pressure. After recombination, the neutral hydrogen no longer feels radiation pressure and can collapse into the gravity wells produced by the dark matter, eventually forming the first stars and galaxies.

4   Stellar winds without metals:
Metals are responsible for the spectral lines which enable UV energy to be efficiently absorbed. Thus momentum is transfered from the radiation field to the gas by means of absorption and scattering. In metal-free gas, there are not strong absorption lines (except those in hydrogen). Thus "line-driven" winds are not expected to exist in the first generation massive stars.

Chapter 17

1   Searching for signals:
The data being collected by some of the world's radio telescopes are being continually searched for potential artificial signals. This analysis is being partially conducted by a public-domain screen saver available from SETI@home project (go to http://setiathome.ssl.berkeley.edu/ for further details). This project represents one of the largest distributed computing efforts in the world.

Index

absorption lines, 94, 270
accretion
    giant planet formation, 147–8, *149*
    mechanism for T Tauri stars, 96–7, 98
    rates in molecular clouds, 70–1
accretion disk *See* circumstellar disks
active galaxies, 237, 285
adaptive optics, 25, 262
Alpha Centauri, *11*, 64
Alpha Persei cluster, 215, 282, 283
ambipolar diffusion, 55, 266
Andromeda Nebula, 10, *221*, 232
angular measurements, 260
angular momentum, 266
    distribution in the Solar System, 143
    gas near the galactic center, 225
    orbital migration and, 278
    solar nebula particles, 146
    transfer in disks, 273
angular momentum problem, 56, 91
arcseconds, 260
asteroids, 166, *167*
astrobiology, 247
Astronomical Unit (AU), 7, 256, 279
atmospheric effects, 19, 25, 46–7, 260, 262
atomic hydrogen
    distribution, 34, *35*, 213–14, 283
    primordial star-forming clouds and, 239–40
    radio emissions, 283
atomic nuclei, 8, 256

Baade's Hole, 216, 283, *284*
Barnard objects, *33, 34, 38*
Betelgeuse (Alpha Orionis), 111, 125, 136
Big Bang, evidence for, 10
binary star systems, 62–9
    birth and youth of, 64–8, *99*, 270
    close binary formation, 68–9
    supernovae among, 137
    white dwarfs in, 202
bipolar phenomena, 236–7, 271
    extragalactic jets, 236
    Herbig–Haro objects as, 75, *78*
    magnetic fields and, 272
    molecular outflows as, 80
birth line, HR diagram, 61, 267
black body radiation, 238, 264, 286

black holes, 23, 207, 261, 270
    Milky Way central region, 227–8
    supermassive black holes, 235, *236*, 244, 285
    type Ib supernovae and, 274, 282
blue supergiants, 263, 269
Bok globules, 36, *38*, 75, 125
brown dwarfs, 6, 59, 180

cannibalization of galaxies, 218
cannibalization of protostars, 123, 128
capture theory, binary star formation, 67
captured satellites, 152
carbon burning in stars, 203
carbon compounds *See* organic chemistry
carbon monoxide, 262–3
    emissions, 80, *81*, 85
    in molecular clouds, 29–30, 34, *36, 37*
carbonaceous chondrites, 164
carbon–nitrogen–oxygen cycle, 242, 258, 279–80
Cas–Tau Group, 215–16, 282, 283
cavitation in molecular clouds, 81, *83, 84*
CCDs (charge–coupled devices), 19, 27, 189–90, 260
central dominant elliptical galaxies, 218, 284
charged particle acceleration by magnetic fields, 91
chondrules, 164–5, 276
circumplanetary disks, 152
circumstellar disks, 94–7
    See also proplyds; solar nebula
    angular momentum problem, 56
    binary star systems, 65, *66, 99*
    clusters of stars and, 179–80
    debris disks and, 160
    dissipation caused by stellar UV, 108
    images of, *58*, 81, *83*
    lifetime, 101–2
    magnetic fields, 90
    rotation, 58–9, 96
    self–luminosity, 98
Class 0 objects, 57–8, 61, 85, 100
Class I objects, 58, 61, 100
Class II objects, 100
Class III objects, 101, 105
clusters of galaxies, 228–9, 245
clusters of stars, 109–21
    circumstellar disk disruption, 179–80
    distances between stars, 180

molecular clouds (*cont.*)

dispersal of young stars from, 93, 109, 112, 273

dust in, 32, 125

dynamical friction, 233

evolution, 36-8, 42, *44*

examples, *49*, 99, *100*, *101*

HAeBe stars and, 106, *107*

heating, 266

with Herbig–Haro objects, *73*

ionization by massive stars, 203

star formation in, 41-2, 48-52

Orion A and B clouds, 128-30

primordial star-forming clouds and, 239-40

radio investigation, 30

rotation, 56

size and structure, 36, 40

star-forming galaxies, 219

in the sub-millimeter region, *47*

molecular hydrogen

carbon monoxide as tracer, 29-30, *36*, 262-3, 283

dissociation in protostars, 57

emissions caused by outflows, 83, *86*

explosion remnants in Orion, 128, *132*

molecular outflows, 80, *81*, *82*

Moon, 152, 154-5

Mt Wilson Observatory,*17*, *18*

multiple star systems, 64-5, 68-9, 71

neutron stars, 269, 270

See also pulsars

formation, 206-7

type II supernovae and, 274, 282

neutrons, 8, 256, 280

novae, 202

nuclear binding energy, 206, 281

nuclear reactions *See* thermonuclear fusion

nuclear starbursts, 233-5

OB associations, 109-12

Solar System origins, 184

fossil associations, 215, 282

Gould's Belt and, 216

prospects for planet formation in, 172, 181

runaway stars, 136

superbubbles from, 209-10, 212

triggered star formation, 133

observatory sites, 17-22

Omega Centauri, 117, 118

Oort Cloud, 158

open clusters, 112-16, *113*, *115*

NGC 3603, 137, *138*

destruction of, 119-20

optical thickness, 57

orbital eccentricity and collision, 147, 155

orbital migration, 183, 278

orbital motion

Milky Way stars, 220, 227-8

pressure gradients and, 146

ordinary matter, 7, 255, 284

organic chemistry, 164, 246, 248

Orion

See also Horsehead Nebula; Trapezium

Herbig–Haro objects in, 72, *73*, 77

views of the constellation, *129*, *130*, *131*, 275

Orion 1a, 1b and 1c subgroups, 133-4, 276

Orion molecular cloud, *47*, *110*, 127-30, 275

Orion Molecular Cloud 1 (OMC1), 89, 180

Orion Nebula, 10, *104*, 172, *173*, *174*, *175*

Orion star forming region, 125, 133

Orion superbubble (Orion's Cloak), 134-5

outflows from galaxies, 234-6

outflows from young stars, 72-93

mechanisms, 89-92

as model for galactic outflows, 237

Overwhelmingly Large (OWL) Telescope, 24-5

oxygen as indicator of life, 249

parallax measurements, 7

parsec, 256

particle formation, solar nebula, 146

Phoebe, 152, *153*

photography, contribution to astronomy, 4, 19

planetary nebulae, 120-1, 200, *201*, *202*, 275

planetary system formation, 143-60

circumstellar disk mass and, 95

Class III objects, 101-2, 105

environment effects on, 183

hazards, 172-84

impossibility for the first stars, 241

meteorites as evidence of, 161, 164

planetisimals, 146-7, *148*

captured satellites as, 152, 276

evidence of heating, 277

Kuiper Belt as a reservoir of, 167

metal cores, 166, 181

meteorite formation from, 165-6

planets

See also extra-solar planets

characteristics favouring emergence of life, 247

composition and history, 145

orbital migration of, 183, 278

plasma clouds, 172, 174, *177*

Pleiades cluster, 112, *113*, *114*

pre-main sequence stars, 61, 97

See also T Tauri variables

pressure, gas, 264

pressure gradients, orbiting particles and, 146

primordial plasma, opacity, 238

primordial star-forming clouds, 239-41

propagating star formation, 133-4, 222, 225

proplyds, 172-84

contamination by supernovae, 181

disk dissipation, 182

evolution, 178-9

planetary formation in, 181-2

silhouetted, 174, *178*

structure, 176, 277, *278*

proton–proton chain, 242, 279-80

spiral galaxies
  emergence of, 244
  spiral nebulae identification as, 217
  star formation in, 219–25
Spitzer Space Telescope, 49–51, *52*, 75–6, *77*
star counts, molecular clouds, 32, 34
star formation
  See also molecular clouds; starbursts
  binary star systems, 65, 67
  colliding galaxies, 229
  concentration in spiral arms, 220–1
  detection, 26
  emergence of life and, 246
  50–100 million year cycle, 199, 216
  first stars, 241–3
  modeling, 53, 55–6, 28, 53
  ordinary matter and, 7
  in other galaxies, 217–37, 245
  propagating star formation, 133–4, 222, 225
  timescales, 58, 60, 243
  triggered star formation, 133
star formation efficiency, 112–14, 271, 273–4
starbursts, 117–19, 231, 233–4
star-forming regions
  infrared image of, *52*
  locations of largest, 118, 137, *138*
  Orion region, 125
  radio and infrared map, *211*
  Sagittarius B1 and B2, 225
stars
  See also interstellar distances
  chemical composition, 6
  colors, 4, 8, 32
  distances of closest, 64
  energy source, 7
  spectral types, 273
star-spots, 103
stellar associations *See* clusters; OB associations
stellar evolution, 9, 269
  low-mass stars, 199–202
  massive stars, 202–7
stellar masses
  See also intermediate mass; low-mass; massive
    stars
  binary star systems, 64
  competitive accretion and, 70–1
  distribution of, 59
  end-states and, 269
  mass loss, 207, 263–4, 280–1
  originating cloud features and, 53, 55
stellar wind bubbles, 203, *204*
stellar winds, 7, 263–4, 280–1
  as dust source, 32
  FUor eruptions, 98
  metal–free stars, 287
  protostellar outflows and, 90
strong nuclear force, 8, 256–7
sub–millimeter region, 47, 51, *55*, 57
Sun
  compared with average stars, 6
  corona and solar wind, 89–90

galactic position, 31, 220, 283
  gravitational effects of Jupiter, 186
sunspots, *102, 103*
superbubbles, 134–5, 210–12, *224*
  Cas–Tau Group, 215
  differential galactic rotation, 222
  fragmentation of super–rings, 216, 283
  in the early Universe, 242, 244, 245
  molecular cloud creation, 214
supergiant phases, 9, 204
  See also massive stars
  blue supergiants, 263, 269
  mass loss from, 281, 263–4
  Orion supergiants, 125–7
  red supergiant phase, 203, 269
supermassive black holes, 235, *236*, 244,
    285
supermassive stars, 7, 255
supernova remnants, 207–9, 211
supernovae, 28, 206–7
  classification, 269–70, 274, 281–2
  frequency, 209, 234
  globular clusters and, 117
  mass qualification for, 282
  molecular cloud destruction by, 114, 116
  open clusters and, 120
  planet formation and, 181
  Solar System contamination by, 166, 181,
    277–8
  supergiant phase preceding, 281
  triggered star formation, 133
  Type 1a as "standard candles", 255
  white dwarfs and Type 1a, 202
supershells, 211, 212, 215
surface temperature, relation to luminosity, 60–1,
    267–9
synchrotron radiation, 208, *209*, 264, 286

T Tauri variables, 93–6, 100–3
  causes of variability, 102–3
  HAeBe stars and, 106, *107*, 108
  prediction of disks around, 174–6
  self–irradiation effects, 179
  Solar System origins and, 145
technology, and astronomical progress, 16, 193–6,
    251
telescopes, 16–27, 260
  ground–based, 24, 28, 189–90
  infrared astronomy and, *46*
  mounting systems, 24, 261–2
  pioneers', 3, 4, 16
  space–based, 25–7
  three functions of, 18–19
temperature scale, Kelvin, 259
  See also surface temperature
terrestrial exo–planets, 187, 190, 192
Terrestrial Planet Finder (TPF), 194, 279
thermonuclear fusion, 8–9
  in early stars, 242
  essential characteristic of stars, 60
  heavy element production, 203, 206, 207